Mushroom Grower

Rajesh Abraham
Amrutha K
Rakesh Menon

2026
Agri Horti Press
New Delhi-110 084

ISBN: 978-93-49690-99-8 (Hardbound)

Published By

Agri Horti Press

New Delhi

E-mail: agrihortipress@gmail.com

Digitally Printed at : Replika Press Pvt. Ltd.

Preface

Mushroom cultivation has evolved from a niche agricultural practice to a thriving global industry, offering diverse opportunities for both small-scale farmers and large-scale producers. Mushrooms are not only valued for their unique flavors and nutritional benefits, but they also play an essential role in sustainability, offering eco-friendly solutions for food production and waste management. As the demand for organic and functional foods continues to rise, mushroom growing has become a key area of interest for agriculture enthusiasts, entrepreneurs, and those seeking alternative crops.

This book, titled *Mushroom Grower*, aims to provide an in-depth guide to the art and science of mushroom cultivation. Whether you're a beginner looking to start a mushroom farm or an experienced grower seeking to refine your techniques, this book offers comprehensive insights into the entire process of mushroom production. It is designed for farmers, agricultural students, hobbyists, and anyone with an interest in growing mushrooms for food, medicinal purposes, or commercial use.

The book begins by covering the fundamentals of mushroom biology, discussing the different types of mushrooms, their lifecycle, and the environmental conditions necessary for their growth. Detailed chapters follow on setting up a mushroom growing environment, including considerations on substrate preparation, inoculation, temperature, humidity, and light.

Special attention is given to the various species of edible and medicinal mushrooms, such as Agaricus, Shiitake, Oyster, and Reishi, along with the specific growing techniques for each type. Readers will also find practical guidance on harvesting, post-harvest handling, storage, and marketing strategies for those looking to turn their mushroom-growing hobby into a profitable venture.

Furthermore, the book explores the benefits of integrating mushroom cultivation into sustainable farming practices. From composting agricultural waste to reducing the carbon footprint, mushrooms offer a unique solution to some of the most pressing environmental challenges of our time.

Through a combination of practical tips, technical insights, and case studies, *Mushroom Grower* serves as both a learning tool and a reference manual. We hope it empowers readers to embark on their mushroom-growing journey, fostering a deeper appreciation for this versatile and valuable crop.

Rajesh Abraham

Amrutha K

Rakesh Menon

Contents

Unit 1

Introduction and Requirements for Growing Mushrooms

1.1 Mushrooms- An gribusiness

Mushroom as a nutritious food, needs to be popularised. Mushroom cultivation will help in eliminating protein malnutrition among people, primarily dependent on cereals, and offer remunerative employment opportunities. As India's share in global mushroom trade is minimal, coordinated efforts among R&D establishments are needed to disseminate technology to grassroots.

Mushroom farming has become an alternative for vegetable farmers as it offers them new business opportunities. Mushroom cultivation can yield quick money if the farmer understands the practical details of the modern techniques and type of species that are in high demands.

Mushroom cultivation is one of the most profitable agribusiness that can be started with minimum investment and space. Mushroom farming in India is growing steadily as an alternative source of income.

1.2 Status of Mushroom cultivation in India

1.2.1 Introduction

Mushroom belonging to fungi species, is a nutritious vegetarian delicacy and a good source of high quality protein. There are presently 3 varieties of mushroom cultivated:

- white mushroom (*Agaricus bisporus*),
- paddy-straw mushroom (*Volvariella vovvacea*)
- oyster mushroom (*Pleurotus sajor-caju*).

Mushroom contains many vitamins and minerals, like B- Complex and iron, and is good source of quality proteins like lysine. Mushroom is completely fat free and also rich in anti–oxidants.

In India, mushroom has been a non-traditional cash crop grown indoors, both as a seasonal crop and under the controlled environmental conditions. Button mushroom is cultivated in temperate regions of India. Oyster, milky, paddy straw mushroom is cultivated in the tropical and sub-tropical regions. Two to three crops of button mushroom can be harvested per year under controlled conditions, while for seasonal button mushroom ,one crop is harvested per year.

Mushroom cultivation in India is more encouraged by government as government provides subsidies to farmers for mushroom cultivation. Since mushroom cultivation business does not create any type of pollution, so due to this reason, large numbers of growers are following this mushroom farming business & earns millions of money from this agricultural business plan.

Cultivation of mushroom in homestead is becoming very popular because of high yield and the remunerative price that it fetches. At present there are mushroom farmers all over Meghalaya identified as trained growers, and training sessions are being held regularly in villages in all the districts of the State.

In India, mushrooms can be cultivated throughout the years in some regions such as Jammu & Kashmir, Himachal Pradesh and in some hilly area of Tamil Nadu, Uttar Pradesh & Eastern states where atmosphere temperature always remains below than 35 'C. In those low temperate areas, two to three crops can be easily obtained throughout the year.

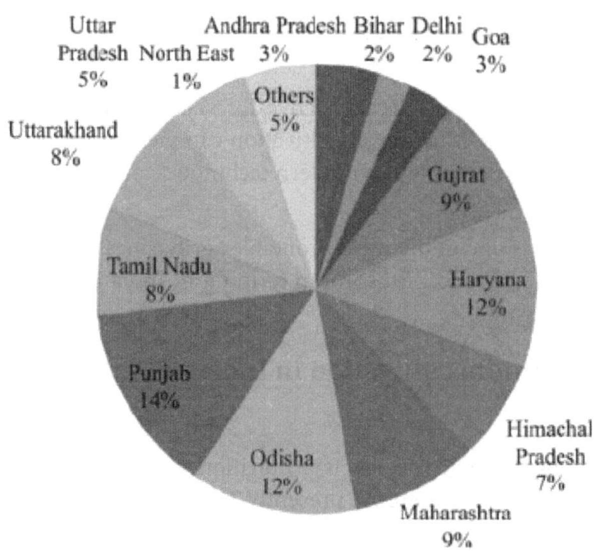

Meghalaya offers great potential for production of Mushroom. A Regional Centre for Training and Production of Mushroom in North Eastern Region was established in 1982 at Agricultural Complex, in Shillong whose aim involves in:

- Production and supply of quality spawn.
- Provide training facilities to interested entrepreneurs for mushroom cultivation.
- Preparation of sterilised compost.
- Dehydration and marketing facilities.

Major export destinations for Indian mushroom are US, Israel and Mexico. India exports mushroom in two forms-fresh and prepared/processed. Button mushroom accounts for approximately 95 per cent of total mushroom exports. Indian exports have been subjected to non-tariff barriers and thus the export trend is fluctuating. Although the current share of India in world exports is less than 1%, India has a great export potential.

Mushroom is efficient means for conversion of agricultural wastes into valuable protein and presents huge potential for generating additional income and employment. In India, the full potential of mushroom cultivation is yet to be unleashed. The cultivation practice has centered on white button mushroom, accounting for 85% of the total production.

The global mushroom industry has expanded very rapidly in the last two decades by the addition of newer types of mushrooms for commercial cultivation. However, mushroom asa vegetable is yet to find regular place among the Indian consumers. Despite of favourableagro-climate, abundance of agro wastes, relatively low-cost labour and a rich fungal biodiversity, India has witnessed a lukewarm response in its growth. At present, the total mushroom production in India is approximately 0.13 million tons.

Even as the mushroom production and consumption are on the rise in rest of the world, India witnesses a luke warm response in its growth. Mushroom industry in India is over whelmingly focused on white button mushroom which is a highly sophisticated and capital-intensive activity.

From 2010-2017, mushroom industry in India has registered an average growth rate of 4.3% per annum. Out of total mushroom produced, white button mushroom share is 73% followed by oyster mushroom (16%), paddy straw mushroom (7%) and milky mushroom (3%). Compared to other vegetables; per capita consumption of mushrooms in India is meagre and data indicates it is less than 100 grams per year. In the year 2016-2017, Indian mushroom industry generated revenue of Rs. 7282.26 lacs by exporting 1054 quintals of white button mushroom in canned and frozen form.

White button mushroom production was confined to temperate hilly regions of India. However, with the development of short method of composting and optimization of fruiting conditions using the chilling system, there has been a remarkable change in its production scenario and spread to all the corners of the country.

1.2.2 Growth of mushroom production in India

There has been significant increase in production of mushrooms in the last few years, especially of the oyster and paddy straw mushrooms in India. The country's production in 2010 was 1.00 lakh metric tons, of which button mushroom accounted for 89% of the total production, followed by oyster (6%), milky (1%) and others (4%). Punjab, Uttrakhand, Haryana, Uttar Pradesh and Tamilnadu were the leading producers of the mushroom in the country at the time of 2010. The present production status revealed that, Maharashtra and Odisha are emerging as the leading states in mushroom production.

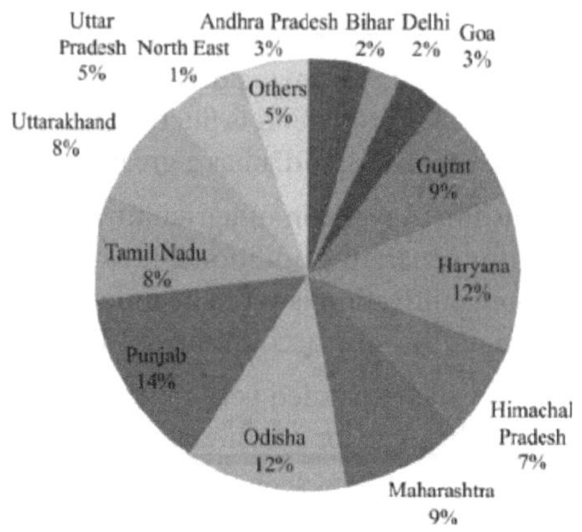

By considering the present production data, mushroom industry in India recorded an average annual growth rate of 4.3%. During this period, the productivity has raised from 20% to 24.5% by the releasing of improved strains in commercial edible mushroom.

In India, there are five mushroom species:

- white button mushroom
- oyster
- paddy straw
- milky
- shiitake

Even though, cultivation technologies of many exotic were standardized, the commercial markets are still dominated by *Agaricus bisporus, Pleurotus* spp. and *Volvariella volvacea.* These three mushrooms are contributing about 96% of total mushroom produced in India. Milky mushroom is indigenous tropical mushroom of the country.

1.2.3 Present scenario

Since 1983, large number of mushroom growers started mushroom cultivation during winter around Delhi, Chandigarh and some districts of Haryana and Punjab.

Button mushroom, straw mushroom and oyster mushroom are the three major types of mushrooms in India used for cultivation. Paddy straw mushrooms can grow in temperatures ranging from 35^0 to 40^0C. Button mushrooms grow during winter. Oyster mushrooms are grown in the northern plains. All the three mushrooms of commercial importance are grown by different techniques. They are grown on special beds called compost beds. Learn how to cultivate mushroom of each types.

Since mushrooms have a very short life, it should reach to consumers within a short time or immediately canned. This will lead to proper marketing of mushrooms. In India, over-production and improper care for marketing have resulted in the increase in price. The cost of production in India at present is comparatively higher than other countries and hence cannot compete in international market.

The yield obtained so far is low due to:

(a) improper infrastructure of preparation of pasteurized compost,

(b) use of ordinary buildings lacking proper temperature control as cropping rooms,

(c) use of low yielding strains,

(d) inadequate supply of quality spawn,

(e) lack of trained manpower,

(f) inadequate research support

The technology for cultivation and processing of mushroom has been developed at CFTRI (Hyderabad), RRL (Jammu) and NBRI (Lucknow). RRL and NBRI are distributing mushroom spawns in rural areas for mass cultivation. Out of these, CFTRI develops technique for processing and drying of mushrooms.

1.3 Status of mushroom cultivation in Haryana

1.3.1 Introduction

Indian agriculture is known for its multi-functionalities of providing employment, livelihood, food, nutritional and ecological securities. Mushroom production leads to bioconversion of agro-residues into nutritious food. India produces about 600 million tonnes of agricultural byproducts, which can profitably be utilized for the cultivation of mushrooms. Currently, India uses 0.04% of these residues for producing around 1.2 lakh tons of mushrooms of which 85% is button mushroom.

India contributes about 3% of total world button mushroom production. In wake of increasing population, increase in awareness about health benefits of mushroom and changing food habits, the demand for various mushrooms is likely to increase. Mushroom cultivation contributes in achieving nutritional and social security along with promotion of value addition, canning and export industries also. Some of mushrooms like oyster, paddy straw and milky mushroom have simple cultivation technology. The cultivation methods for such mushroom varieties suits to Haryana conditions which have also been standardized. Mushroom being rich in protein is considered important particularly for people of Haryana since majority of its population is vegetarian. The cultivation of mushroom has been adopted in the recent years by the farmers of Haryana because the economic returns are much higher compared to other agriculture crops.

The production results are based on different studies of seasonal mushroom farms undertaken in Haryana. About 34.9% of total investment is on fixed inputs while 65.1% is on variable inputs. In fixed investment, major share is spent on construction of mushroom sheds while the variable costs involves in labour (28.09%) followed by that on straw (14.96%) and bran (8.96%). The total returns per kg of mushroom produced was 9.50 on an average selling price of 30.00 per kg. Since benefit cost ratio over variable cost, total production cost and total cost were greater than the mushroom production is a highly profitable proposition. Mushroom produced in state is sold in different markets through commission agents and also sold to processing firms or traders.

The marketing cost of mushroom is quite high and costs like washing, weighing, packing etc have necessarily to be made by growers/farmers. The major marketing costs are commission charges (41.05%) and transportation (36.52%) and such charges are rationalized by collective efforts of farmers and government. Mushroom growers also pointed out the unauthorized charging of commission fee for selling their mushroom from designated farmers sheds where they are not supposed to pay any commission charges. Direct marketing of mushroom by farmers to processing firms and traders also involves manipulative and unethical practices by these like reduction in predetermined contract price on flimsy excuses of quality and delay / even default in payments. When prices are low, they do not purchase from the contracted farmers. Keeping in view, the farmers vulnerability, they have to put joint or group efforts and these may or may not be formal but should be the practical ones.

1.3.2 Cultivation

Button mushroom

- ❖ Button mushroom compost is prepared using variety of base materials like wheat straw, paddy straw, sugarcane bagasse, chicken manure, cakes and brans, etc as per availability in different regions.
- ❖ In major parts of Haryana wheat straw is widely used whereas in some parts, paddy straw is used which is available in abundance.
- ❖ Poultry manure is an good component which meets the nitrogen needs, throughout the state and is used in short method of composting.
- ❖ Growers following long method of composting use chemical fertilizers in place of poultry manure for balancing nitrogen requirement in compost.
- ❖ Poultry manure is a carrier of nematodes and competitor moulds and hence, is not preferred for unpasteurized compost.
- ❖ Haryana farmers have successfully adopted low cost technologies developed by Haryana Agriculture University for seasonal cultivation and contribute over 10,000 tons of mushrooms annually.

Oyster mushroom

- ❖ Cultivation of oyster mushroom (Dhingri) in Haryana has tremendous potential, which is yet to be fully harnessed.
- ❖ Oyster mushroom cultivation technology is very simple as no special compost preparation and casing is needed.
- ❖ Oyster mushroom production is more than the white button mushroom as it can be consumed fresh as well as can be sun dried for future use.

Milky mushroom

- ❖ Milky mushroom, due to its robust size, milky white colour, flavour and long shelf life has attracted attention of both consumers and prospective growers.
- ❖ Milky mushroom nutritive value is also at par with other edible mushroom.

- The cultivation method is similar as that of oyster mushroom, except that it requires casing.
- For promotion of these, efficient low-cost region-specific growing systems and species for seasonal and round- the-year commercial cultivation should be developed and use of spent substrate for recycling as manure or cattle feed may be promoted.

1.3.3 Subsidy on Mushroom cultivation

The Haryana government has launched scheme for promoting mushroom cultivation in four districts of state where an amount of Rs 50 lakh has been earmarked to provide subsidy under the scheme during the current financial year. A 50% subsidy would be given up to 1000 trays per beneficiary. In Haryana, four districts in which the scheme would be implemented includes Kaithal, Jind, Rewari and Mohindergarh, where such districts were not covered under National Horticulture Mission for promoting Mushroom cultivation.

The scheme envisages too give self-employment to small and marginal growers/farmers, landless and unemployment graduates to improve their socio-economic status and to promote ancillary industry like processing and packaging industry. The scheme aim at provide protein foods to overcome deficiency in masses which was mostly vegetarian. The Haryana state was a leading producer of mushroom in the country, growing button mushrooms during the winter season and white milky mushrooms oyster during the summer season.

1.4 Nutritional and medicinal value of mushrooms

1.4.1 Introduction

Mushrooms are classified as vegetables in the food world, but they are not technically plants. They belong to the fungi kingdom. Although they are not vegetables, mushrooms provide several important nutrients.

Mushrooms contain more protein than fruits & vegetable and, Mushrooms can also be low in cholesterol. Apart from their protein content, mushrooms can also be high in certain vitamins like B, C, vitamin D, riboflavin, thiamine nicotinic acid. Also an excellent source of iron, Potassium, and potassium along with folic acid, a component known for improving the blood and avoidance deficiencies.

1.4.2 Nutritional values of Mushrooms

- Indian diet is primarily based on cereals (wheat, rice and maize), which is deficient in protein.
- Supplementation of mushroom recipe in Indian diet will bridge protein gap and improve general health of socio-economically backward communities.
- Earlier mushrooms were considered as expensive vegetable and were preferred by affluent peoples for culinary purposes.
- Currently common populace also considers mushroom as a quality food due to its health benefits.

- Mushroom is considered to be a complete, health food and suitable for all age groups, child to aged people.

Nutritional value of mushroom is affected by:

- species
- stage of development
- environmental conditions

Mushrooms are rich in protein, dietary fiber, vitamins and minerals. The digestible carbohydrate profile of mushroom includes:

- starches
- pentoses
- hexoses
- disaccharides
- amino sugars
- sugar alcohols
- sugar acids

The total carbohydrate content in mushroom varies from 26-82% on dry weight basis in different mushrooms. The crude fibre composition of mushroom consists of partially digestible polysaccharides and chitin. Edible mushrooms commonly have insignificant lipid level with higher proportion of polyunsaturated fatty acids which results in low calorific yield from mushroom foods.

Mushrooms do not have cholesterol, but they have ergosterol that acts as precursor for Vitamin D synthesis in human body. The protein content of edible mushrooms is high, but varies greatly. The crude protein content of mushrooms varied from 12–35% depending on species. The free amino acids composition differs widely and are rich in theronine and valine but deficient in sulphur with amino acids.

Mushrooms contains 80%-90% of water, and 8% to 10% of fiber. It is an excellent source of vitamins especially C and B (Folic acid, Thiamine, Riboflavine and Niacin). Minerals like potassium, sodium and phosphorous are higher in fruit bodies of mushroom and contains essential minerals (Cu, Zn, Mg) in traces but deficient in iron and calcium.

1.4.3 Medicinal values

Edible fungi have revered for their immense health benefits which mainly used in folk medicine. Specific biochemical compounds in mushrooms are responsible for improving human health in many ways which includes compounds like:

- polysaccharides
- tri-terpenoids
- low molecular weight proteins
- glycoprotins
- immunomodulating

The health benefits of mushrooms includes:

- promotion of immune function
- boosting of health
- lowering the risk of cancer
- inhibiting tumor growth
- balancing of blood sugar
- ward off viruses, bacteria and fungi
- reducing inflammation
- support body's detoxification mechanisms

Increasing recognition of mushrooms in complementing conventional medicines is also well known for fighting many diseases.

1.4.4 Health Benefits of Mushroom

Mushrooms have been used for medicinal purposes for thousands of years, to strengthen immunity and preventing bacteria from disrupting the digestion. Mushrooms have been used for thousands of years in traditional asian medicine, and scientists today are beginning to discover that many edible mushrooms are potent disease fighters, which can target malignant or cancerous cells while protecting and even supporting the healthy cells that are already in your body.

1. Good for heart

- The edible mushrooms have little fat with higher proportion of unsaturated fatty acids and absence of cholesterol and consequently it is the relevant choice for heart patients and treating cardiovascular diseases.
- Minimal sodium with rich potassium in mushroom enhances salt balance and maintaining blood circulation in human.
- They are suitable for people suffering from high blood pressure.
- Regular consumption of mushrooms like Lentinula, Pleurotus spp were stern to decrease cholesterol levels.

2. Low calorie food

- The diabetic patients choose mushroom as an ideal food due to its low calorific value, no starch, and little fat and sugars.
- The lean proteins present in mushrooms help to burn cholesterol in the body which makes it as most preferable food for people who strive shedding their extra weight.

3. Prevents cancer

- Compounds restricting tumor activity are found in some mushrooms but only a limited number have undergone clinical trials.
- All forms of edible mushrooms, and white button mushrooms, can prevent prostate and breast cancer.

- Fresh mushrooms are capable of arresting the action of 5-alpha-reductase and aromatase, chemicals which is responsible for growth of cancerous tumors.
- Polysaccharide-K drug is isolated from Trametes versicolor, which is used as leading cancer drug.
- Many mushroom are derived polysaccharides which have the ability to reduce side effects of radiotherapy and chemotherapy.

4. Anti-aging property

- The polysaccharides from mushrooms are potent scavengers of super oxide free radicals.
- In mushroom, antioxidants prevent the action of free radicals in the body, and reduces aging process.
- Ergothioneine is antioxidant found in Flammulina velutipes and Agaricus bisporus which is required for healthy eyes, kidney, bone marrow, liver and skin.

5. Regulates digestive system

- The fermentable fiber and oligosaccharide from mushrooms acts as prebiotics in intestine which carries useful bacteria in colon.
- Fermentable fiber assists the digestion process and healthy functioning of bowel system.

6. Strengthens immunity

- Mushrooms are capable of strengthening immune system.
- A diverse collection of polysaccharides and minerals, isolated from mushroom is responsible for up-regulating the immune system.
- Such compounds potentiate the host's innate and acquired immune responses and activate all kinds of immune cells.
- Macrophages are a type of immune white blood cell that engulfs and digests cellular debris, foreign substances, microbes, cancer cells, and anything else in your body that does not appear healthy to the immune system.

1.5 Benefits of different mushrooms

1.5.1 Introduction

Mushrooms belong to family of Agaricus, which are not plants, but are classified as vegetables. They belong to fungi kingdom and contains host of essential nutrients. They are saprophytes organisms that depends on dead and decaying animal and plant matter for their nutrition. Their body is spread over larger area and is hardly noticed. The part of mushroom we see is only fruit of vegetable. There are thousands of varieties mushroom that forms fungi across the world.

Mushrooms are good natural sources of Vitamin D which contains germanium, selenium, copper, niacin, potassium and phosphorus. They are rich in Vitamin C, protein, calcium and iron and contains insoluble chitin and soluble beta glucans that are good for health.

Nutritional facts Per 100 Grams

22	0.3 g	5mg	318 mg
Calories	Total Fat	Sodium	Potassium
3.3 g	3.1 g		
Total Carbohydrate	Protein		

Vitamins and Minerals

0.01	3 %	2 %	5 %
Vitamin D	Vitamin C	Iron	Vitamin B-6
2 %			
Magnesium			

1.5.2 Types

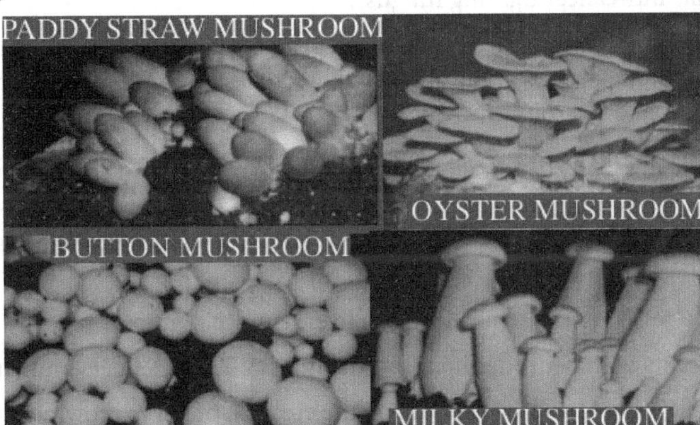

There is various type of edible mushroom available in the world but in India mostly four type mushroom cultivated:

- ❖ White Button Mushroom
- ❖ Portobello Mushroom
- ❖ Oyster Mushroom
- ❖ Paddy Straw Mushroom

1.5.3 Health Benefits of Mushroom

Eating mushroom is beneficial in many ways as mushrooms are high in nutritional value. There are some of the health benefits in all the mushrooms which includes:

- ❖ It helps in building strong immune system

- It helps in prevention of heart diseases
- It helps in preventing cancers
- It helps in weight Loss
- It is beneficial for diabetic patients
- It increases metabolism
- It prevents stomach disease
- It helps in increasing the hemoglobin level in the body

1.5.4 Benefits of different Mushroom Species

Reishi

- It is the species of mushroom which looks like a big white and brown flower made of wood.
- It is loaded with ganoderic acid that helps in reducing cholesterol and also in lowering high blood pressure.
- It has strong immune boosting powers and is revered as an antibacterial, antiviral, anti-fungal, anti-cancer fighting fungus.
- It contained unique polysaccharides that promote longevity by boosting immune system function and preventing abnormal blood vessel formations that could lead to life-threatening cancerous growths.
- In this, triterpinoid ganoderic acid acts to inhibit the development and metastasis of tumors.
- This mushroom species could seek out and eradicate existing cancerous cells in the body.
- It is used to reverse the chemical-driven liver damage in mice.
- It help patients with neurodegenerative disorders like Alzheimers and Huntington's.
- It helps in fighting cancer by supporting the creation of nerve growth factor, which is a protein that is fundamental for healthy neurological function.

Shiitake

- This specie is an excellent source of Vitamin D as it contains natural anti-tumor compound called lentinan.
- It reduces inflammation in the body.
- By using dry shiitake mushrooms and testing with blood before and after will result in better-functioning gamma delta T-cells and reductions in inflammatory proteins.

Oyster

- It is a fungi which grow readily around the world.
- These are rich in proteins, as it contains vitamins B and are cholesterol free.
- These have significant levels of cholesterol-lowering molecule lovastatin.

- In study on HIV patients, it is seen that mushrooms had some anti-retroviral-induced hyperlipidemia.
- With an extraction of two molecular mechanisms from alcohol, the oyster mushrooms inhibits growth of colon and breast cancer cells without significant effect on normal cells.
- These have potential therapeutic/preventive effect on breast and colon cancer.

Porcini

- It is a sort of meaty mushroom which is reminiscent to Portabello which is anti-inflammatory.
- It is a fungus which has compound ergosterol that is capable of cytotoxicity which attack enemy cells.

White Button

- Button mushrooms have grown wild and are eaten by humans since times of early hunter-gatherers.
- As per traditional civilizations, it was known that this mushrooms had special powers as it could grant immortality where pharaohs were deemed worthy of eating or in touching.

Maitake

- This mushroom species has anticancer, antiviral, and immune-system enhancing effects and may also help in controlling high blood pressure and blood sugar levels.
- Such mushroom extract couple which proves highly beneficial to people suffering from insulin resistance and diabetes.
- By consuming half cup per day of maitake mushrooms, it will lead to sweeping of system, in finding abnormal cells and cause them to self-destruct.
- In this, trumpeting bushels triggers the body to release immune system cells that attack and kill malignant cells.

Shimeji

- Such mushroom type has always been used as a defense against asthma.
- It has high potent levels of beta-glucans that helps in suppressing allergic reactions like recurring asthma.
- In this, beta-glucans which boosts the natural healing capability of immune system is a compound that is successful remedy for retarding and destroying growing tumors.

Chanterelle

- Chanterelle mushrooms has highest natural concentrations of vitamin B, especially vitamins B1, B2, B3, and B5.
- Such vitamins play a fundamental role in converting food into energy and keeping the nervous system healthy.

- ❖ These mushrooms are high in fiber, which helps in keeping regular bowel movements and in checking gut health.
- ❖ This single mini-trumpet mushroom has been associated with anti-microbial, bacterial and fungal properties due to high vitamins C, D and potassium.

Black Truffle

- ❖ It has good flavor and has good taste.
- ❖ Such mushrooms contain bliss molecule which is similar to tetrahydrocannabinol, or THC, found in marijuana, which is a chemical that is responsible for many marijuana's mood-enhancing and psychological effects.
- ❖ Expensive black truffles produces compound called anandamide, which triggers the discharge of chemicals and enhances the mood, changes appetite, affects memory and reduces feelings of depression and pain of human brain.

Chaga

- ❖ It is used in form of tea due to its health benefits.
- ❖ It is quiet popular in western hemisphere since it has tested cells that were pre-treated which on extraction with H_20_2 induces oxidative stress.
- ❖ The pretreated cells displayed less damage than cells that did not receive the chaga extract.
- ❖ This mushroom species contains massive amounts of natural black pigment known as melanin, which has high antioxidant levels due to presence of polyphenols.
- ❖ It has highest ORAC score of any superfood.
- ❖ It helps in diabetes, cardiovascular health, immune health, DNA damage protection, and cholesterol.

1.6 Poisonous Mushrooms

1.6.1 Introduction

Edible, inedible and poisonous mushrooms all grow in forest. They are fruiting bodies of variety of fleshy fungi that are found worldwide and appear suddenly from growing vegetative (mycelium) portions containing spores as reproductive units. Fungi lack chlorophyll; their nutritional requirements are met by utilizing organic material from a saprophytic, parasitic, and/or mycorrhizal life cycle.

1.6.2 Poisonous Mushroom Identification

Mushrooms with white gills are often poisonous, so are those with a ring around the stem and those with volva as volva is often underground, which is important to dig around the base of a mushroom to look for it.

Mushrooms with red color on the cap or at the stem are either poisonous or strongly hallucinogenic. The most notorious red-colored mushroom is Amanita muscaria, which has been consumed for thousands of years to produce visions.

1.6.3 Common Poisonous mushrooms

There are many groups of common mushrooms that have poisonous species among their ranks where first group Amanitas, are some of the most deadly mushrooms of all. There are a few edible amanita species, though it is best for amateur mushroom hunters to avoid amanitas altogether which has following features:

- all amanitas have pale gills (white or yellow) and white spores.
- starts initially by universal veil (often egg shaped).
- when mushroom emerges, veil breaks and forms volva and some often remains on caps as patch or series of warts/flakes.
- many amanitas have partial veil that covers gills early on and once broken, forms a ring on upper stalk
- gills free from stalk or attached, but never running down it

Galerinas are group of small, brown mushrooms that are challenging to identify. It is the most important species to get to known is deadly galerina as it contains dangerous amatoxins, and could be deadly if consumed. To identify it, look for these features to identify these poisonous mushrooms:

- Cap is small between 1-2" across, and brown, yellowish or tan in color
- Cap is smooth, sticky when moist
- Gills are tan or brown, and spores are rusty-brown or brown
- Stalk is skinny, often less than 1/4" and without scales
- Veil present in the beginning, then forms a small ring on upper stalk which can disappear with age/weathering
- This species is found of growing on decaying wood and woodchips

- Often grows in small groups or alone, but not in big clusters

The false morels and close relatives are other dangerous group, and its good to know how to identify these poisonous mushrooms with following features:

- Caps are strange, brain-like, with very wrinkled surface when mature
- Do not have honeycombed with pits like true morels
- Cap color is brown, reddish-brown or yellow-brown
- Cap attached to stalk at multiple points, but attached everywhere
- Stalk is usually smooth or slightly grooved, not folded in a complicated pattern inside
- Grows in the springtime on ground or rotten wood in forests

Of the less dangerous species of poisonous mushrooms, one of the most common gastrointestinal irritants is emetic russula, which has following features that will help in identification:

- The cap of such medium sized mushroom is bright red to pink and smooth
- Gills are white, stalk is white, spores are white
- Stalk without thicker base, column-like and snaps cleanly like a piece of chalk
- Entire mushroom brittle
- No veil, ring or volva
- Taste is extremely hot and peppery

1.6.4 Toxic compounds contained by Mushrooms

Amatoxins (Amanita-toxins)

These are poisonous compounds carried by the Amanita family, as well as members of the galerinas, conocybes and lepiotas. Here is a short list of some of species that contain these toxins:

- Destroying Angel (Amanita ocreata)
- Fool's mushroom (Amanita verna)
- European Destroying Angel (Amanita virosa)
- Death Angel (Amanita bisporigera)
- Deadly Conocybe (Conocybe filaris)
- Deadly Parasol (Lepiota josserandii)
- Deadly Galerina (Galerina autumnalis)

Amatoxin poisoning is extremely dangerous and has a high fatality rate. The trouble with them is that they often cause liver and kidney failure. The symptoms are also often delayed and may not be visible for 6 hours or up to 24 hours after ingestion.

Symptoms include violent vomiting, bloody diarrhea, severe cramps. Often, these symptoms may subside and appear to indicate recovery. Lastly, relapse occurs and failure of kidneys and the liver leads to death.

There is no known true antidote for amatoxins. Recent hope has been shown in the use of milk thistle extract given via injection can slow or stop symptoms of amatoxins. However, limited studies have been done to show the effectiveness of this medicine and it is not yet widely used for treatment against mushroom poisonings.

Gyromitrin

Gyromitrin or (monomethyhydrazine) is a very toxic compound which is found in false morel (Gyromitra esculenta) and hooded false morel (Gyromitra infula). The former species is responsible for a significant amount of mushroom poisonings. There appears to be wide variations in toxicity in different geographical strains. Other related species are:

- ❖ *Gyromitra californica*
- ❖ *Gyromitra caroliniana*
- ❖ *Gyromitra fastigiata*
- ❖ *Gyromitra gigas*
- ❖ *Verpa bohemica*
- ❖ *Helvella sp.*

Though some the species on this list are eaten by mushroom hunters after thorough cooking, it is not recommended. Symptoms appear 2 hours to up to 1 day after ingestion. They may include diarrhea, vomiting, gastro-intestinal distress and headaches. In some cases, kidneys, liver and even the red blood cells are damaged which may result in death.

Gastrointestinal Irritants

This is the largest group of poisonous mushrooms. Thankfully, the effects of mushrooms in this group are temporary and much milder than the first 2 groups mentioned. Here are some of the common species in this group:

- ❖ Flat-topped Agaricus (Agaricus praeclaresquamosus)
- ❖ Yellow Stainer (Agaricus xanthodermus)
- ❖ Mock Meadow Mushroom (Agaricus californicus)
- ❖ Felt-ringed Agaricus (Agaricus hondensis)
- ❖ Satan's Boletus (Boletus satanas)
- ❖ Red-pored boletus (Boletus pulcherrimus)
- ❖ Slender red-pored Boletus (Boletus erythropus)
- ❖ Green-spored Parasol (Chlorophyllum molybdites)
- ❖ Emetic russula (Russula emetic)
- ❖ Fragrant russula (Russula fragrantissima)
- ❖ False Chanterelle (Hygrophoropsis aurantiaca) pictured below
- ❖ Sulfur Tuft (Hypholoma fasciculare)

The symptoms of gastrointestinal poisoning are experienced within 4 hours or as soon as 20 minutes. They include nausea, cramps, diarrhea and vomiting. Thankfully, these generally pass once the irritant is expelled from the body. Occasionally, severe cases require hospitalization.

Muscarine

This toxin contained in some poisonous mushrooms has effects on the parasympathetic nervous system. Some of the common poisonous mushrooms that contain muscarine are:

- Lurid Boletus (Boletus luridus)
- Jack O'Lantern Mushroom (Omphalotus olivascens)
- Sweat Mushroom (Clitocybe dealbata)
- Conic Fiberhead Mushroom (Inocybe fastigiata)
- Fluff fiberhead mushroom (Inocybe lanuginose)

Effect of consuming poisonous mushrooms containing muscarine may include excessive salivation, perspiration, tears, severe vomiting and diarrhea. They might also include visual disturbances, difficulty breathing, a drop in blood pressure and an irregular heartbeat. Symptoms generally dissipate within 24 hours. Though rare, death may result from respiratory failure.

Coprine

These are compounds contained by members of the inky cap group (Coprinaceae). Though the substance itself is not toxic, it has some disturbing symptoms when mixed with alcohol. Here are some of the members in that group reported to have caused poisoning:

- Inky Cap (Coprinus atramentarius)
- Coprinus variegates

The danger of coprine occurs when mushrooms that contain this compound are eaten before, during or after the consumption of alcohol. Coprine when mixed with alcohol in the human body can create symptoms such as rapid heartbeat, light headedness, a metallic taste, reddening of nose and eyes, nausea and occasional vomiting. Generally recovery is complete and within 24 hours.

Ibotenic Acid

This acid is converted into the compound muscimol in the body, after it is consumed. There are 3 main poisonous mushroom species that contain this toxin in our region:

- Fly Amanita (Amanita muscaria)
- Panther Amanita (Amanita pantherina)
- Gemmed Amanita (Amanita gemmata)

Muscimol is an intoxicating and psycho-active compound. Some of the symptoms appear about 30 minutes or up to 2 hours after consumption and may include confusion, mild euphoria, loss of muscular coordination, chills, profuse sweating, visual distortions, a feeling of increased strength, delusions, convulsions and hallucinations. Many consumers of this mushroom become drowsy and fall into a deep slumber, awakening later with no memory of their experience. The amount of ibotenic acid present in these species of mushrooms varies not just from species to species, but from individual mushroom to individual mushroom. The effects also vary greatly from person to pers

The use of Amanita muscaria by humans has been recorded for thousands of years in Eurasia, especially in the far North such as in Siberia. Shamans in some of those same areas today continue to consume this species in order to experience altered states of consciousness and use them in healing ceremonies for other members of their tribe or family.

Psilocybin & Psilocin

Perhaps some of the most famous mushrooms are those containing the chemicals psilocybin and psilocin. This group of so called, "magic mushrooms" belong mainly to the genus psilocybe and panaeolus. Not all of them stain blue when bruised. Here are some of the most well known species:

- Potent Psilocybe, Wavy Cap (Psilocybe cyanescens)
- Magic Mushroom (Psilocybe cubensis)
- Azur Cap (Psilocybe azurescens)
- Liberty Cap (Psilocybe semilanceata)
- Panaeolus subbalteatus
- Panaeolus cyanscens

Symptoms of ingestion are often similar to those of LSD consumption. These might include heightened color perception, visual distortions, rapidly shifting shapes and images, laughter and elation, and hallucinations or delusions. Nausea, vomiting and anxiety are not unheard of. Symptoms are generally temporary and last about several hours.

The danger in consuming these mushrooms comes often from misidentifying them and consuming a deadly species. To a novice mushroom collector, the wood-loving galerina species might be mistaken for certain psilocybes. Consuming galerina mushrooms can be deadly!

1.7 Composting of button mushroom

1.7.1 Compost

The substrate used for button mushroom is a partially decomposed organic matter prepared under aerobic conditions and is generally termed as compost. In India, straw of wheat or paddy have generally replaced horse manure as the base material for mushroom compost. It is known as synthetic compost. Several formulations of compost have been worked out, the most commonly used are:

Long method compost (unpasteurized)	
Solan	
Wheat straw	300kg
Calcium ammonium nitrate or ammonium sulphate (20.6% N)	9kg
Urea (46% N)	3.6kg

Potassium sulphate or muriate of potash	3.0kg
Wheat bran or	30.0kg
Spent brewer's grains	40.0kg
Gypsum	30.0kg
Nemagon (60%)	40ml
Furadan 3 G	150g
Lindane or BHC 5% dust	250g
Molasses	5kg
Ludhiana	
Wheat straw	300kg
Poultry manure	60kg
Wheat bran	7.5kg
CAN	6kg
Urea	2kg
Superphosphate	2kg
Potassium sulphate	2kg
Gypsum	30kg
Lintox	60ml
Srinagar	
Wheat straw	300kg
Molasses	12kg
Urea	4.5kg
Wheat bran	50kg
Muriate of potash	2kg
Cotton seed meal	5kg
Gypsum	15kg
Lindane	250g
Shillong	
Paddy straw	400kg
Ammonium sulphate	9kg
Urea	3.6kg
Sulphate of potash	3.0kg
Single super phosphate	3.0kg
Wheat bran (or flour)	30.0kg
Gypsum	30.0kg
Molasses	5.0kg
Temik	40.0gm
BHC 5%	250gm
Kelthane or Ecalux	40ml

Any of the above formulations can be used for preparing long method compost (LMC) which is accomplished outdoors in about 28 days. The constituents included in the formulations ensure the N levels initially at 1.5–1.75 and finally at 1.25%, as also the C:N ratio between 25 and 30 at starting and 16–18 at the end.

The straw after thorough wetting for 24 hr is mixed with the bran fertilizer mixture prepared with two-thirds quantities of ammonium sulphate and urea and the entire quantity of SSP and SOP added to 15kg of moistened wheat bran and left overnight covered with wet gunny sheets. The substrate so prepared is formed into a large heap to encourage intense microbial activities causing the generation of heat reaching up to 75°–80°C.

The heap is broken and remade on the sixth day after adding the bran fertilizer mixture made the previous night with the remaining ingredients and slurry made with molasses, nematicide and insecticides in 10 litres of water. Every 3–4 days, the heap is broken and remade (turned) after adding water to maintain around 75% moisture and allowing aerobic conditions. Normally 7–8 turnings are necessary with addition of gypsum at third and BHC or Lindane at the 7th turning. The compost after seventh or eighth turning is ready for seeding (spawning) if free from ammonia, otherwise more turnings are necessary.

Short method compost (pasteurized)	
Solan	
Wheat straw	1000kg
Chicken manure	400kg
Brewer's grain	72kg
Urea	14.5kg
Gypsum	30kg
Bangalore	
Wheat straw	300kg
CAN or ammonium sulphate	9kg
Super phosphate	9kg
Wheat bran	15kg
Gypsum	30kg

Thus short method compost can be prepared accordingly if completed in 18–20 days in outdoor and indoor phases. During the outdoor phase, the straw is pre-wetted and spread before the entire quantities of chicken manure and Brewer's grains are added in layers. Ample water and trampling makes the loose stack almost anaerobic which is normally given a turning after 2–3 days. An aerobic heap is prepared 2 days later after adding the full quota of urea. About 3–4 turnings on alternate days complete the outdoor phase.

The phase-II is done indoors either in a bulk chamber or in a pasteurization room. The bulk chamber is especially designed for phase-II composting and is fitted with boiler-fed steam-pipes and a blower. Bulk pasteurization in such chambers has now become more popular particularly in bigger farms. The phase-I compost is directly filled into the chamber up to a height of 6–7 feet, while it is filled in trays or shelves to a depth of 17–22cm in a pasteurization room. In either case, the temperature is allowed to rise first to 48°–50°C and after 6–8hr it is raised by steam injection to 57°–59°C for effective pasteurization of the compost. When both the air and compost temperatures have reached and maintained at this range for 4–6hr, fresh air is introduced slowly to lower down the compost temperature to 50°–52°C for conditioning, which takes 3–4 days and is indicated by the absence of ammonia. Introduction of more fresh air brings down the temperature of compost to 25°–28°C, when it is ready for seeding.

1.7.2 Spawning and spawn run

The compost made by long or short method is filled in trays, or shelves or more commonly in polybags, after mixing the spawn through compost @ 0.5%. The spawned beds are kept covered with formalin dipped paper sheets or by closing the mouth of the bags,. The beds if incubated at 24°C, with relative humidity maintained between 80–85%, get fully impregnated with mushroom mycelium and when the spawn-run is completed (in 2–3 weeks), the compost turns light brown from deep brown and is ready for casing.

1.7.3 Casing and case run

The spawn run beds are then covered by 1–2" thick layer of casing material, which is necessary to initiate fruiting, though its role in fruiting is only partially understood. Casing material should have the characteristics like poor in nutrient, good water-holding capacity but a texture permitting good aeration and a pH of 7–7.5. The best and the most commonly used casing material is the peat-moss, which is directly used for casing after adding lime or chalk to adjust the pH and also after pasteurization. However, due to its paucity in India, one of the following mixtures is being used in our country.

- 1–2 years old and rotten cowdung + clay loam soil (1:1).
- 1–2 years rotten cowdung + clay loam soil + 2 years old spent compost (1:1:2).
- 2 years old spent compost + sand + lime (4:1:1).
- Garden loam + sand (4:1)

The casing material is pasteurized either by steaming (60°–65°C for 6 hr) or by adding 3% formalin solution (3 litres formalin in 40 litres of water), covering with a polythene sheet for 15–20 days. Before casing, the material is allowed to cool or become free from formaldehyde. To encourage the mushroom mycelium to fully impregnate the casing soil, temperature is maintained around 24°C for 7–10 days. After this the room temperature is brought down to 14°–18°C and ample ventilation is provided to reduce CO_2 level, preferably below 1,000ppm. The relative humidity of the room is maintained between 85 and 90%. The casing layer is given light spray of water to prevent its drying.

1.8 Infrastructure for button mushroom

1.8.1 Introduction

Mushrooms belong to separate group of organisms called Fungi which lacks the green matter (chrolophyll) present in plants and grows on dead and decaying organic materials. From these decaying substrates, they absorb their nutrition using fine thread like structures called as mycelium that penetrates into the substratum and are generally not visible on the surface. After the mycelium has grown profusely and absorbed sufficient food materials, it forms the reproductive structure which generally comes out of the substrate and forms fruiting body, commonly known as mushroom. The mushroom fruiting body may be umbrella like or of various other shapes, size and colour. It consists of cap or pileus and a stalk or stipe but others may have additional structures like a veil or annulus, a cup or volva, performing various functions in the life-cycle of the fungus.

Button Mushroom is most popular mushroom variety grown and consumed across the world. In India, its production earlier was limited to winter season, but with the technology development, these are produced almost throughout the year in small, medium and large farms, adopting different levels of technology. The species being grown in most farms is the white button mushroom belonging to Class Basidiomycetes and Family Agaricaceae.

1.8.2 Components of Mushroom Farm

- ❖ Composting Unit
- ❖ Outdoor Phase-I composting platform/indoor bunkers or aerated chambers
- ❖ Indoor Phase-II in peak heating/bulk past-chamber
- ❖ Peak heating chamber
- ❖ Bulk pasteurization chamber
- ❖ Cooling of compost in summer months a special requirement
- ❖ Casing pasteurization chamber
- ❖ Spawn unit
- ❖ Spawn laboratory
- ❖ Cropping unit
- ❖ Seasonal cropping rooms
- ❖ Environment controlled cropping rooms
- ❖ Environment control, air conditioning and forced air circulation
- ❖ Ancillary units
- ❖ Post harvest handling unit
- ❖ Pre-cooling chamber
- ❖ Canning hall with canning line
- ❖ Packaging room

Composting unit

- ❖ The components of composting unit depend on the method of compost production chosen.
- ❖ In case of indoor compost production, the requirement of composting yard will be reduced and it will be 1/3 of normal yard required.
- ❖ Today, trend is for indoor compost production due to environment legislation where small pre wetting area, and small covered composting yard requires with minimum of two-phase-I bunkers and one phase-II tunnel.

- Size of all these structures depend on the production targets of unit and size and numbers of tunnels.

Pre-wetting area

- This area is constructed closer to the road.
- It is a simple cemented structure having a saucer like depression in the center so that it looks like a lagoon and water remain collected during prewetting of compost ingredients.
- Center of the lagoon is around 1 ft deep.
- In this, extra water of lagoon is collected in goody pit built for the purpose at convenient place around the wetting area for its reuse.
- Floor of wetting area should be such that it can withstand the load of front-end loader while performing the wetting operations.
- The floor is normally not covered and is open to the sky.
- The wetting area terminates in the composting yard.
- In this, water connection with 2”-3” dia. pipe should be available in area permanently with additional portable hosepipe for use during pre-wetting.
- One dewatering pump with hose needs to be installed in goody pit to pump out the run-off water for its reuse during pre-wetting.
- Water in goody pit may be aerated continuously to avoid foul smell.

Composting yard

- The composting yard is required for phase-I of composting which is a prerequisite for short method.
- The composting yard is necessarily covered with shed having 2-3 ft sidewalls on two sides where rain will not interfere in normal process of composting.
- The foundation of composting yard to be laid on a firm ground and should necessarily be reinforced if mechanization of composting yard needs to be done as it has to withstand the load of heavy machines.
- In this, floor is given a run-off of 1 cm per running meter away from the bulk chamber and towards the goody pit end.
- The roof of outdoor composting platform is built on tresses or RCC pillars 16 ft high with GI or any other suitable roofing.
- The composting yard should be big enough to hold maximum compost stacks for phase-I of composting.
- When adopting indoor compost production wetted ingredients are just made up into heap for 3-4 days and do not require rick formation in such a case a small platform can suffice the purpose.

Phase-I tunnels (bunkers)

- This facility is required when indoor composting is employed at the farm.
- These are specially built non-insulated tunnels having full width opening at the front.
- In this, dimension of bunkers depend on output of compost required.
- The bunkers are 1.5 times more the size of the phase -II tunnels.
- It has plenum which is constructed below the actual floor.
- It has a perforated concrete floor with 1 cm openings at a distance of 1ft each to entire floor area which is constructed above the plenum having RCC /steel gratings with 20% opening.
- The entire surface area of tunnel is serviced by centrifugal fan having $1/4^{th}$ capacity of phase two blower where ventilator displacement is of 50 m^3 per hour ton of compost at 50 mm WG water pressure that suffices the purpose.
- In this, plenum floor pressurizes entire airspace beneath the concrete floor, which allows air to move up in substrate by the holes or by series of slates.

Air handling units of the tunnel (AHU)

- For good pasteurization and conditioning of compost in tunnel, specific need of air and ventilation are required, which can be done by providing/installing air handling unit in tunnels.
- In this, effective pasteurization and conditioning is obtained when 150-200 m^3 air per ton of compost per hour is blown by compost mass.
- To have effective pasteurization and conditioning, high speed centrifugal fan is chosen and is placed on slope end of ventilation duct in underground service area.
- Compost is spread over plenum on the grated floor in about 2-2.2 meter thick layers where nylon nets are kept under compost for mechanization.
- Using grated floor and nylon net, there is resistance of around 60-65 mm WG during pasteurization which takes resistance of air ducts, in and out openings, perforated floor, etc.
- Static pressure of blower fan should be around 100 mm WG at 150-200 m3 air per ton of compost per hour.
- The blower fan in this, is well protected internally and should be made up of sheet steel.
- Aluminum is good for air ducts and should at least be 2 mm thick and there should not be any leakage in the duct system.
- In this, ducts are insulated with glass wool or any other suitable material.
- Here fresh air is required in tunnels and as the air is drawn from open atmosphere, chances of fungal spores contamination exists, so incoming air in ventilation duct needs filtration that needs to pass through 2 mm fungal spore filters.
- The pre filters and filters needs washing at regular intervals.

❖ The inlet and exhaust openings to be fitted with flap valve, that gets open when positive pressure is created in the tunnel with its dimensions similar.

Casing pasteurization chamber

❖ Casing pasteurization chamber is a small bulk chamber having all required components that are needed for tunnel.

❖ In this, plenum is not having any slope and capacity of blower for proper steam injection and its uniform distribution in casing mass is around 1/4 the capacity of tunnel.

❖ The size of chamber depends on the size of compost chamber and size of growing rooms.

❖ In this, one chamber load should provide casing for one compost lot from each tunnel.

❖ In this, casing in the chamber can be treated in bulk where it is filled up to height of 90 cm against the tunnel where compost is filled upto the height of 2-2.2 meters.

❖ In this, casing after wetting is filled in perforated wooden/aluminum trays which are stacked one over the other in chamber and steam treated at 65^0C for 6-8 hours.

❖ The chamber can be built near to the composting yard or in composting yard with separate casing mixing platform.

Spawn unit

❖ The layout plan of spawn laboratory should contain total built in area as 60½ x 30½ x 12½ which should be good enough to house the entire spawn unit.

❖ The built up area is divided in different work areas like boiling/autoclaving room, inoculation room, incubation room, cold room, store, office and delivery area.

Cropping unit

❖ As mushrooms are grown indoors under simulated environment which is specially created for its growth, the cropping rooms are required to be built specially for the purpose.

❖ Two types of cropping rooms are built suiting to particular requirement where seasonal growing and environmental controlled growing results all through the year.

Environment controlled cropping rooms

❖ The environment controlled cropping rooms are built like hermetically sealed chambers where the air movement is controlled manually or semi automatically using mechanical control systems.

❖ Such cropping rooms are insulated and its dimensions are determined by the amount of compost to be filled in the room.

❖ Rooms with greater length and narrow width gives better results as far as air handling inside the room is concerned.

- Cropping room with capacity to take compost from one bulk chamber is advantageous as one bulk chamber load can be straightaway filled in one cropping room.
- The cropping cycles to be considered will determine the numbers of growing rooms in the unit.
- Today, 60 days cropping cycle is taken where minimum of six crops are taken / room in a year and minimum of 12 rooms are required to have constant supply of mushrooms from the unit round the year.
- For this, all room is filled with spawned compost every after 5 days.
- In this, bulk chamber and cropping rooms of 20-25 tons compost capacity are considered to be operationally efficient, as filling/emptying operation and spawning can conveniently be done in one day when machines are not to be used.
- In this, bigger units have the growing rooms handling compost to tune of 60 tons or more.

Processing unit

- The design of the processing unit should be such that it should maintain strict hygienic conditions and hence needs special care.
- In this, the floor needs to be well laid off with kota stone having slope at one end with walls having ceramic tiles till the height of 5-6 ft.
- The height of the canning hall should be not less than 14 ft in any case.
- The unit so selected should be clean and away from the composting yard.
- Floor of canning hall should have enough strength to support the weight of different utilities to be installed that are required for canning operation.
- Canning hall should be big enough looking to future needs or processing of items.
- In this, all doors and windows should have wire mesh shutters to prevent the entry of insects and flies.
- The unit room should have exhaust fans at the appropriate places.

1.8.3 Farm Design for Commercial Button Mushroom Cultivation

White button mushroom is a temperate mushroom requiring cooler climate for its growth. It is an indoor crop and is an ideal tool in converting agricultural wastes in to proteinaceous food. In early days its cultivation was mainly confined to the hills. In eighties, growers realized the potential of this crop and started its cultivation in northern plains in winters when climate was suitable for its growth. Many entrepreneurs in the plains ventured and started button mushroom cultivation round the year by employing artificial cooling facilities. Today button mushroom cultivation is done throughout the country. Many growers are cultivating button mushroom seasonally while many preferred for round the year cultivation.

Mushroom being an indoor crop does not require arable land, except for some nonagricultural land to build the infrastructure for preparation of substrate, raising of crop preparation of spawn and postharvest handling. Button mushroom is grown seasonally and in environment controlled cropping houses requiring building of basic infrastructure. In this, seasonal growing is done for 3-4 months when outside temperatures are favourable for the crop, i.e., during winter months in North West plains and from September to April in the hills .

Seasonal cultivators of button mushroom are using traditional methods of its cultivation and are mainly cultivating button mushroom in thatched structures using long method of composting. The growers take single crop in entire season and harvest 12-15 kg mushrooms/100 kg compost. Environment controlled units are cultivating button mushroom round the year by using suitable infrastructure at their disposal that covers modern composting yard with bulk pasteurization facilities. Many units have shifted to indoor composting while new upcoming units have chosen to produce their compost entirely by indoor method. Besides these facilities they are having insulated cropping rooms and other ancillary structures required for mushroom cultivation. Few bigger units are having their own spawn lab and processing unit as well. Suitable infrastructure including different machineries are required at farm to carry out different operations to govern the whole process of cultivation in required fashion so that one gets optimum returns from his/her farm in such competitive environment.

1.8.4 Selection of Site

Before selection of site, the following points have to be taken into consideration for greater operational efficiency and cost effective production of button mushroom at farm, which includes:

- The site selected should be away from municipal limits and grower/entrepreneur should purchase sufficient land in one go looking at future expansion.
- The site selected should be serviced by motorable road, or nearer to road head to reduce costs on transportation of raw materials to farm/finished product to the market.
- Plentiful availability of water at site either through perennial source or should have sufficient underground water.
- Easy availability of raw materials like straw and poultry manure should be made available at chosen site at cheaper rates.
- Availability of cheap labour in abundance.
- Uninterrupted proper power supply at the chosen site.
- Nearness to the market for the proper disposal of the produce.

1.9 Spawning unit

1.9.1 Spawning

As mushroom matures, it produces millions of microscopic spores on mushroom gills lining the underside of mushroom cap. Such spores functions similar to seeds of higher plant. However, growers do not use mushroom spores to seed mushroom compost as they germinate unpredictably and are not reliable. Mycelium propagates vegetatively from germinated spores, which allows spawn makers to multiply the culture for spawn production. Specialized facilities are required to propagate mycelium, so mushroom mycelium remains pure. Mycelium propagated vegetatively on various grains or agars is known as spawn, and commercial mushroom farmers purchase spawn from companies specializing in its manufacture.

Spawn makers start spawn and makes process by sterilizing mixture of millet grain or rye, wheat, and other small grain, water and chalk. Sterilized horse manure formed in blocks was used as the growth medium for spawn up to 1940, known as block or brick spawn, or manure spawn. Once sterilized grain has mycelium added to it, grain and mycelium is shaken 3 times at 4-day intervals over 14-day period of active mycelial growth. Once the grain is colonized by mycelium, the product is called spawn, which can be refrigerated for few months.

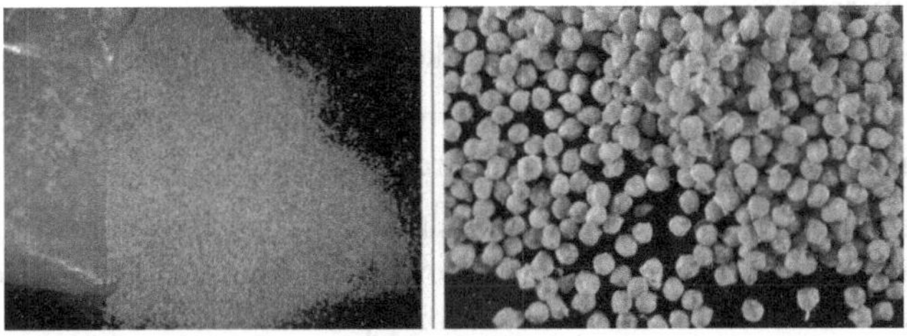

Spawn is distributed on compost and then thoroughly mixed in compost. For years this was done by hand, broadcasting the spawn over surface of compost and ruffling it in with small rake-like tool. In recent years, for bed system, spawn is mixed in compost by special spawning machine that mixes compost and spawn with tines or small finger-like devices. In a tray or batch system, spawn is mixed in compost as it moves along a conveyer belt or while falling from a conveyor in a tray. The spawning rate is expressed as unit or quart per so many square feet of bed surface. These rates are sometimes expressed on the basis of spawn weight versus dry compost weight; a 2 percent spawning rate is desirable.

1.9.2 Spawn unit

The spawn unit set up in particular place carter the needs of mushroom growers in cluster of villages in locality.

The Spawn production unit have mushroom growing unit of its own which test the yield potential of its own spawn and demonstrate the cultivation technology to aspiring growers.

Requirements

Land and building-

The land and building comprises of four rooms that will be used as spawn production unit such as:

- ❖ Inoculation Chamber: It is used for mycelial multiplication/putting mushroom mycelia in sterilized substrate under strict aseptic condition. The important equipment and accessories in this room will be laminar air flow machine and ultraviolet light.
- ❖ Incubation Chamber: It is used for keeping the growing mycelia under required temperature and humidity conditions. In this, inoculated cultures or spawn bottles are kept in racks.

- ❖ Media Room-cum- office room: It is used for preparing media/substrate for preparation of cultures.
- ❖ Store room: It is used for keeping the materials used for spawn production,

Equipments/ Machines/Accessories

- ❖ Autoclave: It is used to sterilize the media in culture tubes and wheat medium in spawn bottle.
- ❖ Laminar Air flow machine: this is used for maintaining aseptic/ germ free condition in the inoculation chamber
- ❖ Refrigerator: It is used to keep the cultures and media for longer period. In this, fresh mushrooms can also be stored.
- ❖ Furniture: It comprises of iron racks for keeping spawn bottles, table with stools
- ❖ Accessories: This will include weighing balance for weighing chemicals, LPG stove with cylinder to boil wheat and to prepare culture media.
- ❖ Glassware and chemicals: It includes culture tubes such as borosil to make culture tubes which will be used to raise and maintain pure mushroom cultures. Glass bottles are used as containers of mushroom spawn.
- ❖ Measuring cylinder: It is used for measuring water during media preparation.
- ❖ Inoculation needle: It is used to transfer mushroom mycelia to culture tubes and spawn substrate
- ❖ Plastic baskets: These are heat resistant plastic baskets/ wire baskets that are used to keep the media having culture tubes in autoclave for sterilization.
- ❖ Dextrose: It is used for preparation of PDA, which is a medium for culturing mushroom mycelia
- ❖ Agar-agar: It is prepared from see weeds, which is used for solidifying the PDA medium.
- ❖ Formaldehyde: It is a solution which is used for fumigation of inoculation and incubation chambers to maintain germ free condition
- ❖ Rectified spirit: It is used for fueling the spirit lamp and sterilizarion of inoculation needle and hands
- ❖ Calcium carbonate: It is used for mixing in boiled wheat grains to increase alkalinity in the culture medium.
- ❖ Wheat: It is used as base material for mushroom spawn
- ❖ Non- absorbent cotton: It is used to plug the mouth of the spawn container.
- ❖ Butter paper and rubber band: It is used to wrap the mouth of spawn bottle using butter paper and is tightly secured by rubber band
- ❖ Glass marking pencil: It is used to mark the mushroom species, date of preparation of spawn on the bottle

Unit 2

Compost Preparations for Button Mushroom

2.1 Long method of composting

Mushroom can be cultivation using compost, spawn and casing mixture having high quality and for having good mushroom yield, good quality compost is required. The material on which mushroom is grown is called compost, which are made with the many articles which are mixed in particular proportion.

There are two method of preparation of compost such as:

- long method
- short method

In both the methods, the compost mixture is made in open on floor by fermentation but in short method compost are prepared in a specific room where the mixture is filled for two weeks which are called chamber or tunnel.

The floor of the chamber is sieve type and below this, an air is passed through the blower that maintains uniform temperature of the compost. As a result of this, compost is prepared early and is prepared by such method about doubled productivity in comparison to compost made by long method.

Long method of composting is the oldest method and now exists only in few pockets of the world mainly because of poor productivity, proneness to attack by the competitors, and also due to more time and labor consuming process. This method is completely an outdoor process and takes about 28 days, though production of long-method compost in lesser duration has also been achieved. But the biomass loss in this process is very high (30-35%) and the quality as well as productivity is poor, besides the environmental problems it creates.

In long method of composting, pasteurization is avoided, which will makes the compost poor in quality and often gives variable yields. Under long method, composting is done on a cement floor in an open or under roof, but sides are to keep open.

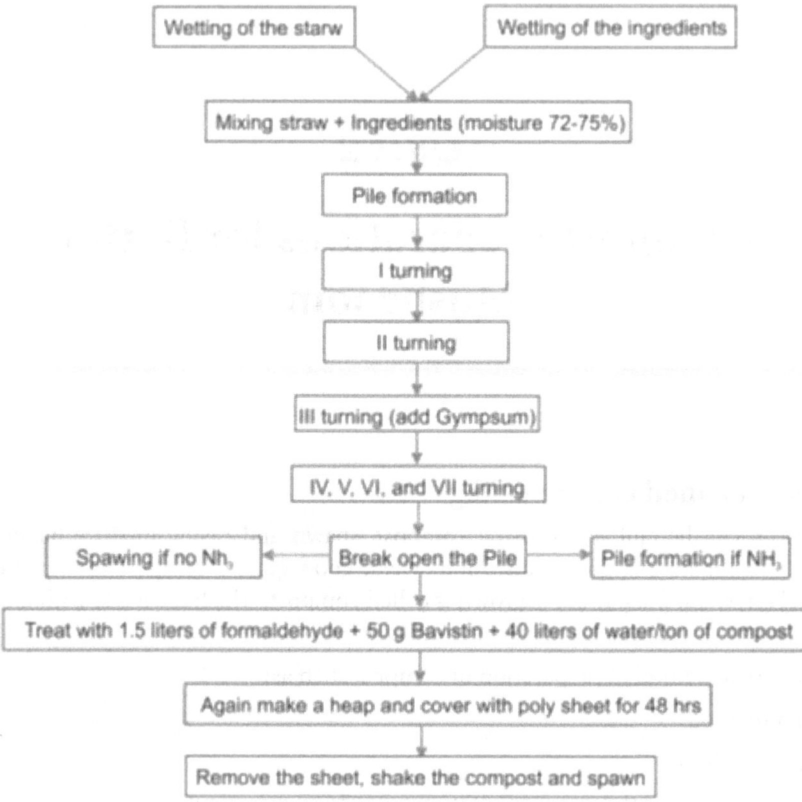

Natural compost:

This compost is traditionally prepared by using horse manure and the barn waste consisting of straw bedding of wheat of barley. These are taken in the proportion of 4:3 and to this mixture, generally 100 kg of chicken manure and 5 kg of urea are added per tonne of substrate. The manure is kept as heap of about one metre height and is regularly examined and turns down repeatedly when it emits ammonia smell. This is o be done for every three – four days. Finally every tone 25-kg of gypsum is added.

Artificial Compost:

Many formulae are available and ingredient to be used varies as per the locality and availability of materials. A widely used formula is given as:

Formula	**Quantity**
Wheat straw (chopped to 8-2- cm)	250 g
Wheat bran	25 kg
Ammonium sulphate of Calcium ammonium nitrate	4 kg
Urea	3 kg
Gypsum	20 kg

Chopped straw is spread over the floor and water is sprinkled thoroughly to wet the straw. Mix all the ingredients except gypsum. Finally the mixture stacked to a height of one metre and compacted using wooden boards. This mixture is turned periodically on 5th, 10th, 14th, 18th, 22nd and 26th day. The gypsum is added in two equal splits on the 14th day and 18th day. It is advisable to add nematicides like nemagon on the 22nd day @ 40-50 ml per tone of substrate. In mushroom houses where insect pests are also a problem, 10-15 ml of malathion is also added during the final turning and sufficient quantity of water is also added on the heap.

Preparation of compost by such method is very old concept and has been abandoned in most parts of the world. Compost prepared by such method besides taking more time gives low yield, as it is prone to attack by many pests and diseases. Yields obtained using such compost range between 10-15kg of mushrooms per 100 kg of compost. However, higher yields to the tune of 18-22 kg have also been reported by some seasonal growers who take single crop in the entire season and thereby extending the harvest period for three to four months depending upon the climatic conditions.

Compost by this method is prepared on clean cemented platform. If such facility is not available then a simple brick platform can be used. Most of the growers who are cultivating this mushroom seasonally do not have any of the above two facilities and are producing the compost in open fields. Composting yard should preferably by covered by G.I. or asbestos roofing. If such facilities do not exist then provision should be there to cover the heap during rains. For a medium size farm producing around 20 tons of compost in one operation, a platform of size 60'x40' (18 m x 12 m) is sufficient enough. As a thumb rule 1.5 tons of compost at start of composting occupies around one square meter of space. Besides composting yard, provision should also be there to store the base materials. In general around 4000 litres of water is required per ton of raw material for its proper wetting. The ingredients of compost are straw, wheat bran, urea etc. Any of the region specific synthetic formulation without chicken manure can be tried for compost production.

Features:

- ❖ In long method of composting, compost is made on cement floor which should be cleaned.
- ❖ The floor can be made in a field which is open or can be a shed.
- ❖ If composting method is chosen to be done on open field, then protection of heap should be done from rains by covering it with water.
- ❖ Chopping of straw into pieces is done and the pieces are chopped with size of 20 cms each.
- ❖ The straw will be spread on the floor and water is sprinkled on it to make it wet.
- ❖ In this, straw is left for an entire night in water, which is filled in the tubs or large containers.
- ❖ Ingredients such as rice bran, poultry manure are mixed with straw which is wet and made into piles of height 2 meters and width 1 meter.
- ❖ In this, heap is made to compress by using little pressure.
- ❖ The ingredients can be added to wet straw while forming a pile in terms of layers.
- ❖ The pile can also be prepared by making use of a wooden mould having three boards made of wood.
- ❖ One wooden pile will be for the end while other two are for the sides.
- ❖ The boards which are at the sides are separated from the end board using clamps.
- ❖ The ingredients of compost that is mixed are kept together in wooden mould and will be compressed in subtle way.
- ❖ The sideboards which have been separated from the end board will move forward as per their lengths.
- ❖ The ingredients in mixed compost are placed in wooden mould and long pile of composts are prepared.
- ❖ If pile is long, pipes which are perforated will be placed in vertical position in the pile for purpose of aeration or circulation of oxygen.
- ❖ It is very important to turn the pile or rotate it numbers of times and should be in such a way that each and every part of compost should come to the middle where decomposition is quick and compost can be heated to maximum for managing diseases caused by pests.
- ❖ After completing each turn, sprinkling of water should be done to replace the loss of water caused by evaporation.
- ❖ When the compost is available for spawning, it should have certain features.
- ❖ The appearance of compost should be light brown.
- ❖ There should not be any smell of ammonia in compost.

- ❖ When the compost is squeezed in subtle way using hand, then moisture which is present in the compost should come out in form of dampness and fragments of compost should get bonded.
- ❖ There should be complete decomposition of compost and Carbon and nitrogen ratio should be 6:2 with contents of moisture in compost to be at 66%.
- ❖ If during composting, moisture is less than required percentage, then action of bacteria gets blocked and escape of nitrogen takes place in form of Ammonia or other forms of it which are volatile.

2.2 Short method of composting

Based upon the observations of Lambert that productive compost came from the regions of the pile having temperature between 50-60°C and adequate supply of oxygen, The short-method of composting was developed mainly because it took lesser time than the long method. The concept and process was indeed a revolution in the cultivation of button mushroom.

The short method of composting mainly consists of two phases: outdoor-composting for 10-12 days (Phase-I) followed by pasteurization and conditioning for 6-7 days inside specialized insulated structures, called tunnels.

It is the process in which the mixture of straw converts into selective medium which is suitable by using composting that can be done in two different stages:

Stage 1

- ❖ This stage will take place after the pre-wetting is done and the mixing of raw material with the straw is done as per the long method of compost.
- ❖ To make the composting process quick, the heap will be introduced.
- ❖ The stacks will be prepared in an open field or in a shed which is open on one side.
- ❖ The material will be rotated a number of times.
- ❖ This stage will take up to 2 weeks for completion after pre-wetting is done.
- ❖ The temperature of the stacks middle portion will be at 67% by this time, which is sufficient for the death of the pests to occur.
- ❖ The layers of the pile which is external will not have that much of temperature.

Stage 2

- ❖ This stage is considered as pasteurization or a peak-heating stage.
- ❖ The process of composting is done in an environment which is controlled and will be continued until the compost is tested in terms of nutrition and can be considered selective for the mushroom growth.
- ❖ This stage takes 12 days for completion.
- ❖ The main thing that differentiates stage 1 and stage 2 is environmental control.
- ❖ The compost conditioning will happen at a temperature of 50°C.

Based upon the temperature conditions maintained inside the tunnel, stage 2 can be divided into two sub-phases:

- pasteurization (57-60°C for 6-8h)
- conditioning (45-48°C for 5-6 days)

Importance of conditioning has been linked with the growth of desirable thermophilic microorganisms; pasteurization and conditioning are essential for getting the selectivity in compost.

2.3 Raw Material and Ingredients for Composting

2.3.1 Base materials

- Base materials are bulk component of compost.
- Various crop residues can be used for such purpose by using wheat straw, which is favoured all over the world.
- Quality compost can be prepared by many other materials like paddy straw, hay, barley, oat, maize stalks and leaves, sugarcane bagasse, sugarcane trashes and leaves, soybean stalks, mustard stalks, etc.
- These materials should preferably be freshly harvested/procured and should be around 5-8 cm long.
- The base materials act as reservoir of cellulose, hemicellulose and lignin which is used by A.bisporus during its growth as carbon source which provides small quantity of nitrogen.
- Besides being nutrient source, base materials add bulk to compost and proper physical structure to substrate which ensures ample aeration during composting for build up of microflora needed for composting process.
- Rice and barley straw are very soft and degrade quickly during composting and absorb more water as compared to wheat straw.
- Using such materials, care must be taken on quantity of water used for wetting, schedule of turnings and adjustment to the rate and type of supplements.

2.3.2 Supplements

- The base materials do not have adequate amount of nitrogen and other nutrients required to start the fermentation process.
- The compounding mixture is supplemented with other materials having nitrogen and carbohydrate sources to get proper CN ratio.

These materials can be classified as follows:

a. Animal manure

- Horse manure is best material for compost preparation and due to difficulties in procuring good quality horse manure, use of such materials has been restricted to few farms.

- More and more farms are switching over to easily accessible materials.
- Chicken manure has proved to be the best alternative to horse manure while other manures viz., pig, cattle and sheep have also been tried for compost preparation but with limited success.
- All these manures provide nitrogen to the compounding mixture, little of carbohydrate is also provided.
- These materials are highly variable in composition and their N-content may vary from 1-4 percent and it is released slowly during composting process.
- In addition to provide nutrients, they increases bulk of compost, which is an important factor under Indian conditions.
- If horse manure is used in composting, it should be used along with bedding and urine, as it may not require any further supplementation.
- If it is not having enough bedding and urine when collected from clean stable, supplementation with inorganic nitrogen along with some wheat straw may prove useful.
- Chicken manure if used, should preferable be a deep litter chicken manure having nitrogen contents above 3%.
- If such manure is not available, then manure from cages can also be used.
- Chicken manure is used under short method of composting, but some of growers are using it under long method and are get good yields, while some have met with failures.
- Chicken manure harbours heavy population of pathogenic nematodes and harmful fungi including Sepedonium maheshwarianum, Stachybotrys atra, Papulaspora sp. and Verticillium sp.
- Growers should avoid to use such material under long method of composting.

b. Carbohydrate sources

- Such materials are required to hasten the composting process, to balance C/N ratio and also for establishment of bacterial flora in compost.
- Molasses, wet brewer's grains, malt sprouts, potato wastes, apple and grape pumice can be employed as carbohydrate source, since these materials provide available nutrients to microorganisms.

Nitrogen fertilizers

- In such fertilizers, urea, calcium ammonium nitrate, ammonium sulfate can be kept.
- Nitrogen content of these fertilizers is very high (24-46%) which is released quickly, resulting in quick establishment of microflora.

Concentrate meals

- In this, animal feeds are considered which include wheat or rice bran, dried brewer's grain, meal/cakes of soybean, cotton seed, castor, sunflower, etc.

- ❖ Such materials supplies nitrogen and carbohydrates which in animal manures, are released slowly.
- ❖ Nitrogen content may vary from 3-8% depending upon the source.

Supplements to rectify mineral deficiencies

- ❖ With carbon and nitrogen, bisporus also needs less amount of potash, phosphorous, calcium and magnesium for its growth.
- ❖ Fertilizers like muriate of potash and superphosphate can be kept in this category.
- ❖ With gypsum and calcium carbonate, gypsum has stabilizing effect on ammonium content.
- ❖ An increased ammonium concentration is obtained with gypsum, which is an indicator of productive compost.
- ❖ Gypsum serves as calcium source for mushroom as it converts oxalic acid into calcium oxalate.
- ❖ Need of phosphorous, potassium, and magnesium is achieved by chicken manure or horse manure when compost is produced by short or by indoor method.
- ❖ Long method compost where chicken manure is not added addition of these materials may be required.

2.4 Pasteurization of compost

2.4.1 Compost

- ❖ The fungi agaricus cannot be sown in ordinary soil. If well fertilized, a higher plant like lettuce will not thrive in dark growing room on mushroom farm.
- ❖ With higher forms of life, mushrooms should find organic matter like protein and fat vital to their life which is ready-made selective substrate called as mushroom growing compost.
- ❖ Various byproducts from other farming operations like wheat straw, cotton seed hulls, dried poultry waste, and proteinaceous meals when blended together, begins with compost process.
- ❖ The mixture is turned regularly to allow homogenization of ingredients, aeration and moisture adjustments in compost heap.
- ❖ Ricks or rows are built by specially designed equipment which allows grower to control moisture and temperature of composting process.
- ❖ It takes around 28-30 days of outdoor composting to prepare for pasteurization process.

2.4.2 Pasteurization

- ❖ The compost is filled in trays using specially designed material handling equipment.
- ❖ It is stacked in pasteurization chambers where final stage of composting takes place with more controlled environment.

- During six days of the phase, compost maintains in very close temperature range to allow adequate growth of beneficial organisms and eradication of antagonistic organisms.
- Steam assists naturally occurring heat of compost to reach temperatures that selectively kills many harmful organisms.
- The successful pasteurization completes the composting process.

2.4.3 Mushroom Compost

- In past, mushroom compost was prepared by mixing ingredients, then periodically turning over the mixture for purposes of aeration as micro-organisms in mixture develop, and then flowing hot steam by mixture to kill or deter further development of certain types of bacteria and other organisms.
- Many variations of the process have been used, but in all cases, compost is treated in relatively small quantities or batches, and it has been difficult to provide standard pasteurized compost from one batch to another.
- It is apparent that unless entire mass of compost material is treated in essentially same manner, quality of compost will vary from place to place in the mass.
- The initial compost mixture comprises of horse manure, hay, gypsum and other ingredients are composted in Phase I compositing process and well mixed, fluffed and placed in pasteurization chamber of novel construction.
- The floor of chamber has perforate central section surrounded by perforate fence spaced wih certain distance from outer walls of chamber.
- The space in the fence is filled with compost mixture.
- A mixture of air and steam is blown in large space below the floor and flows more or less uniformly up by floor and out through entire pile of compost mixture.
- The amounts of steam and fresh air which are introduced in recirculating gas are respectively adjustable to allow operator to maintain optimum growth conditions for development of aerobic thermofils.

2.4.4 Purpose of pasteurization and conditioning

- It reduces the time of composting.
- It converts ammonia into microbial protein most of which otherwise goes waste in the atmosphere in LMC
- It conditions or sweetens the compost under definite set of temperature and aeration, uniformly making compost more selective for the growth of A.bisporus.
- It kills or inactivates insects/pests/diseases and competitors of A.bisporus which, if present hamper the growth of A.bisporus thereby reducing the yield.
- Conditioning increases the biomass of thermophilic organisms specially that Scytalidium thermophilum, which later on is utilized by the mushroom mycelium as food

- Through conditioning more compost per unit weight of ingredients is produced
- Conditioning and pasteurization increases the yield of mushrooms.

2.5 Benefits of Short method

Short method of composting has many advantages over the long method:

- more compost per unit weight of the ingredients,
- higher productivity of mushrooms,
- less chances for pests and diseases,
- shorter duration
- less environment pollution

Unit 3

Spawn Production for Different Kind of Mushrooms

3.1 Different type of media preparation for raising cultures

3.1.1 Introduction

Mushrooms can be grown on small scale with correct growing medium, environment and conditions and if a grower cannot find mushroom-growing medium at supply store, he/she can make his/her medium with few common gardening supplies. Mushrooms prefer loose compost mixture that holds enough water for growth without causing rot. The right ratio of ingredients in the growing medium provides all the nutrients needed for healthy mushroom growth.

There are two ways of raising pure culture and they are:

- tissue culture
- spore culture

Mushroom culture can be taken from spores or from tissue. In germinating spores, many strains are formed, which are compatible with one another, while some are not. In taking a tissue culture from living mushroom, the grower preserves the exact genetic character of the contributing mushroom. With spores, a single strain must be selected from the multitude of strains created. In both cases, the result is a network of cells known as mushroom mycelium.

Mushrooms reproduce through spores where chances of mushroom spores germinating and producing a mushroom are less. In laboratory, isolated from airborne contamination, the probability of success is much improved.

Tissue culture technique is used to bring the edible mushroom to pure culture so that the mushroom fungus can further be used to prepare spawn, which is essential material for mushroom cultivation. In this, mycelium covers the entire surface in a week's time and culture becomes ready for further multiplication.

Raising of Culture

For raising the culture, a pure strain has to be developed, and then to increase the mycelial mass which can be done by initially growing the mycelium on enriched agar media in a dish and then on grain or sawdust/bran. On flat pane of dish, contaminants like molds and bacteria become readily apparent as it is easy to see if mushroom mycelium is pure and free of contamination, experienced cultivators propagate mycelium in dishes and then inoculate grain or sawdust/bran that has been sterilized in jars.

When such grain or sawdust filled jars G1 have grown by mushroom mycelium, known as spawn, which can be either used to inoculate other 10-20 more grain-filled jars, G2 or inoculate bulk substrates like straw, wood, or compost. Spawn G1 masters are best grown in regular mouth quart mason jars while G2 spawn is grows out in mouth half gallon and/or gallon jars. Also, G3 spawn can be created from spawn G2, if required and no extra expansions from grain-to-grain transfers to be made beyond G3 as contamination occurs and not be detected until it is too late.

The liquid culture allows a grower to use small mycelial culture from single dish to inoculate hundreds of grain jars in fraction of time it takes with such method. Preferences vary with every cultivator, so mushroom tissue culture is highly individualized art, but FP promotes liquid culture as improvement over more labor intensive, traditional methods.

With many species, grain spawn can be laid in trays, cased with moisture-laden soil-like layer, and fruited. Once tissue culture is mastered, this is easy way to grow mushrooms. As biomass of mushroom mycelium exponentially multiplied from small fragment of mycelium, sterility of the laboratory is of paramount importance. Micron filters solves contamination problem in lab and pay more for themselves by considering contamination they prevent and cultures/time they save.

3.1.2 Culture media preparation

Culture media are essential for isolation of mushroom fungi and maintain them in pure culture in test tubes or in petri plates. The fungal cultures may be grown in liquid media or on solid agar media.

The pure cultures are raised on convenient culture medium which are solidified state due to the addition of agar-agar. The culture of cultivated mushroom can be raised by using following media:

Wheat extract agar medium

- Boil 32-g wheat grains with 1 litre of distilled water for 2 hours and filter after 24 h.
- Add 20g agar to 1 litre of filtrate and boil.
- pH of medium is maintained at 6.5.
- Fill about 5ml of medium in each test tube.
- Autoclave test tubes at 15 lb. psi for 20 to 30 minutes

Lambert's agar medium

- Add 10 g glucose, 0.5 g magnesium sulphate, 1.9 g potassium dihydrogen phosphate and 20 g agar to one litre of distilled water.

- pH of medium is maintained at 6.5.
- Fill about 5ml of medium in each test tube.
- Autoclave test tubes at 15 lb. psi for 20 to 30 minutes

Malt extract agar medium

- Add 20 g malt extract and 20 g agar to one litre of distilled water and boil for 5 minutes.
- pH of medium is maintained at 6.5.
- Fill about 5ml of medium in each test tube.
- Autoclave test tubes at 15 lb. psi for 20 to 30 minutes

Compost extract agar medium

- Boil 200 g ready synthetic compost with 2 litres of distilled water for about 2 h and filter after 24 h.
- Add 35 g agar to 2 litres of filtrate.
- Adjust pH of medium at 6.5-7.0.
- Fill about 5 ml medium in each test tube and sterilize at 20 lb. psi for 1 h.

Potato dextrose agar medium

- Boil 250 g peeled potatoes in one litre of distilled water till these become soft and filter.
- Add 20 g dextrose and 20 g agar to the filtrate and raise volume of water to 1 litre.
- pH of medium is maintained at 6.5.
- Fill about 5ml of medium in each test tube.
- Autoclave test tubes at 15 lb. psi for 20 to 30 minutes

3.2 Instruments required for spawn Production

3.2.1 Spawn

Spawn is the vegetative mycelium from selected mushroom grown on convenient medium like wheat, pearl millet, sorghum, etc for raising mushroom crop. It involves in preparation of pure culture of mushroom from tissues/spores that is maintained on any agar medium, with culturing on sterilized grains and on multiplied grains.

3.2.2 Spawn laboratory

The spawn comprises of mycelium of mushroom and supporting medium which provides nutrition to the fungus during its growth. While exact area of spawn laboratory depends on many factors, with area around 90x30x13'. The lab has different work-areas like cooking/ autoclaving room, inoculation room, incubation room insulated with air-conditioning, washing area, store office and one cold room heavily insulated for storage of spawn.

3.2.3 Equipments in spawn laboratory

- ❖ Two steam autoclaves electrically or steam operated.
- ❖ One boiling vessel,
- ❖ Laminar flow
- ❖ One refrigerator
- ❖ One BOD incubator
- ❖ One pH meter
- ❖ Glassware
- ❖ Chemicals
- ❖ Non-absorbent cotton
- ❖ Polypropylene bags or bottles
- ❖ Air conditioning equipment's
- ❖ Steel racks
- ❖ Exhaust fans
- ❖ Filters

The minimal requirements for a spawn production unit are:

- ❖ sterilization unit (pressure cooker, autoclave)
- ❖ sterile environment: inoculation box or laminar airflow cabin
- ❖ laboratory equipment like Petri dishes, test tubes, scales, alcohol, flame
- ❖ incubation room
- ❖ ingredients for media preparation
- ❖ substrate material (grain, wooden sticks (skewers), sawdust, or even oil palm fruit fibre)

- ❖ pure culture or fresh mushroom of the desired mushroom species strain
- ❖ spawn containers (such as bottles or plastic bags)

3.3 Methods of sterilization

3.3.1 Introduction

Mushroom requires moist environment and acceptable growth medium to grow. When growing mushroom, care must be taken to avoid contamination which could kill the mushrooms or lead to sickness, so to prevent contamination, sterilization of mushroom growing medium is required.

Contamination of mushroom growing medium takes place in many ways, but the most common contaminants are yeast cultures and bacteria, as other chemical or biological contaminants also affects mushroom growth. Contamination reduces nutrients available for mushrooms or breaking down of mushrooms so that they can decay. If a grower grows edible mushrooms, bacterial growth results in sickness if mushroom grows in contaminated growth medium.

3.3.2 Sterilizing Growth Media

Sterilizing of growth medium kills biological contaminants before mushrooms begin to grow. Different sterilization methods are available as per personal preference such as:

- ❖ type of growth medium used
- ❖ amount of sterilize required

Grower can sterilize the growth medium by pressure cooker, microwave, peroxide or by cooking it at required temperature. Any equipment like trays or humidity-controlled containers which are used by the grower should be sterilized.

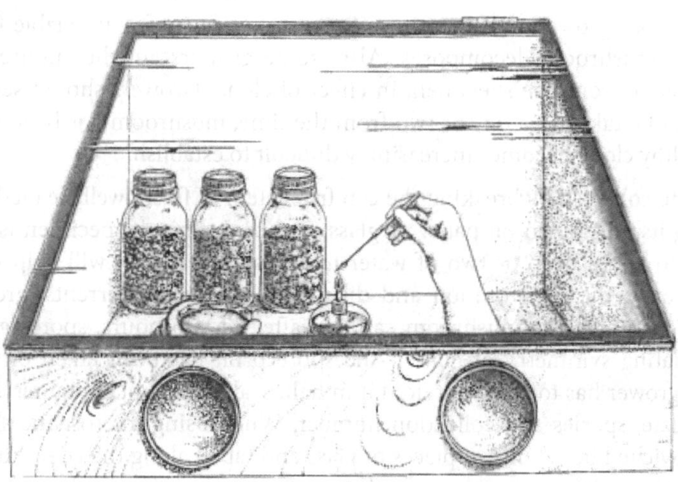

Sterilization reduces the likelihood of growth problems or illness caused by contamination as some mushrooms do not grow in sterilized growth media and instead must be grown outdoors or in nutrient-rich environments that do not lend themselves to sterilization.

3.3.3 Different sterilization methods

There are five different sterilization methods which are used for sterilization of substrates such as:

T1- Hot water treatment:

The substrates were first soaked in cool water for 12 hours, taken out form water after soaking. Water was boiled up to 850C, soaked straw was put in the hot water and boiled for 30 min

T2- Autoclaving:

The substrates were first soaked in cool water for 12 hours, taken out form water, excess water drained out, filled in the polypropylene bags and the polypropylene bags filled with moist substrate were autoclaved for 20 mins. at 15 lbs pressure.

T3- Formaldehyde treatment:

Paddy straw was treated by soaking in formaldehyde solution (50ml/ ltr.water) for 12 hours.

T4- Bavistin treatment:

Paddy straw was treated by soaking in bavistin solution (2g/ ltr.water) for 12 hours.

T5- Treatment with ordinary water (control):

The substrates were first soaked in cool water for 12 hours.

3.4 Single spore and Multi spore cultures

3.4.1 Introduction

Many mushroom growers start a culture from spores as spores are viable for weeks to months after mushroom decomposes. Also, some growers cut the culture into pieces of interior tissue from live specimen, in effect of clone. Growers should see that tissue cultures must be taken in a day or two from the time, mushroom has been picked, after which a healthy clone becomes increasingly difficult to establish.

Spores can be collected by breaking the cap from stem of fresh, well cleaned mushroom and keeping its gills down on paper or glass surface. In this, if specimen is dried, then grower has to add a drop or two of water to cap surface which will help in release of spores. To lower the evaporation and disturbance from air currents, grower should keep a cup/glass over the mushroom cap and after certain hours, spores gets fallen as per the radiating symmetry of gills. If the spore print has been taken on paper, then mushroom grower has to cut it out, fold it in half, seal in airtight container and label the print with date, species and collection number. When using microscope slides, spores can be sandwiched between two pieces of glass and taped along the edges to prevent the entry of contaminant spores. A spore print carelessly taken or stored can easily become contaminated, decreasing the chance of acquiring a pure culture.

Multispore culture

After getting spore print, mushroom culture begins where mushroom grower should sterilize the inoculating loop or scalpel by holding it on flame of alcohol lamp/butane torch for 5-10 seconds till it gets red hot. Then mushroom growers has to cool down the tip by inserting it in sterile media in dish and scrape some spores off the print and then transfer the spores by streaking the tip of transfer tool across agar surface. The method calls for scraping of spore print above an opened dish allows to free fall on the medium. When beginning with new culture from spores, it is best to inoculate at least three media dishes to improve the chances of getting a successful germination. Mycelium started in this manner is called a multispore culture.

3.5 Ingredients of spawn production of button mushroom

Button mushroom is also called white buttom mushroom or champignon. It is a secondary decomposer which cannot feed on fresh substrate but instead, substrate must be decomposed first.

In spawn-production, mycelium from mushroom culture is kept on steam-sterilized grain, and in time the mycelium completely grows through the grain, such grain/mycelium mixture is known as spawn while spawn is used to seed mushroom compost.

Spawn production ingredients

1. Wheat grains were used as support media for spawn production which can be boiled for 30 min for softness.
2. Remove extra water and allow grains to cool.
3. Mix water with calcium carbonate and calcium sulphate powder with 10g/kg of wheat grains.
4. Use glucose bottles and fill these grains with 250 g/bottle.
5. Wrap the bottle with non-absorbent cotton, paper and then autoclaved at 15 psi for 30 min.
6. Once autoclave bottles are cooled, shake it to restore transparent visibility of glass and then inoculate aseptically with mycelial bits from slant cultures.
7. Keep the bits in such a manner that their mycelium touched the grains.
8. Incubate bottle at 22±1°C for 15-20 days.
9. Culture bottles produces Master/mother cultures, so after 10 days, the culture was shaken so that grains were transferred in 10-15 freshly prepared bottles.
10. Spawn bottles were stored at 25^0 C to use for spawning the substrate.

3.6 Isolation of pure culture of different mushrooms

3.6.1 Introduction

Culture media are mixtures of nutrients or substances used in the laboratory for cultivation of many microorganisms. Fungi may be cultured on a variety of media types which can be either solid or liquid depending on experiment. Most fungi grows in medium with high carbohydrate source, nitrogen source, pH of 5-6, and temperature range from 15-37 C. As no medium will sustain growth of every fungal species, type of medium on which fungus grows will vary from species to species. Some fungi like downy mildews cannot be cultured in vitro, and survive only on plant host.

There are two general types of culture media:

- natural
- synthetic

Natural media are made of natural substrates like herbaceous or woody stems, seeds, leaves, corn meal, wheat germ and oat meal. The exact nature of such substrates varies from time to time and cannot be duplicated exactly. This media is usually easy to prepare but they have disadvantage that their composition is largely unknown such as corn meal agar, potato dextrose agar, V-8 juice agar, and dung agar.

Synthetic media, has ingredients of known composition and such media can be duplicated with exactness each time they are made and contain defined amounts of carbohydrates, nitrogen, and vitamin sources. Czapek-Dox medium, glucose-asparagine and Neurospora crassa minimal medium would fall in this category.

3.6.2 Isolation

Hyphal Tipping

To separate and isolate one species of fungus from an undesired contaminants, hyphal tipping is usually performed which is also used when isolating fungal species from infected plant tissues. For this, grower requires:

- Dissecting scope
- Scalpel knife
- Contaminated plates
- Lots of patience

With dissecting microscope, growers will see the species of interest at high magnification and locate each strands of hyphae of fungus. After flaming the scalpel, cut the end of hyphal strand of about 1 mm from end and take small piece of agar where hyphal end is located and transfer to another agar plate. Many hyphal tips are required to ensure that grower have isolated the fungus of interest.

Single Spore Isolation

To isolate pure culture of particular species, single spore isolation is applied which can be used to remove individual spores from species of interest. Such procedure can only be used if fungus sporulates in culture or in planta, or to use hyphal tipping. Depending on what grower does, there are two ways to isolate single spores. When working with plant material, it is possible to observe each spores from infection lesion on plant using dissecting microscope by dipping needle in sterile distilled water to wet the end and removing single spore from infected tissue and then placing the spore on agar medium.

If spores are small and difficult to manipulate or having contaminating fungi, it may be necessary to remove several spores in bulk and dilute them in sterile distilled water. Even though there may be other contaminating fungi along with fungus, the idea is to dilute them away from each other. It will then be possible to isolate the fungi from each other either by hyphal tipping or by single spore isolation without worry of nearby contaminants.

3.7 Production of mother spawn/master spawn

3.7.1 Mother Spawn

Mushroom spawn is a substance that has inoculated with mycelium, which serves as vegetative growth of fungus. Mycelium is a thread-like collection of cells which is mushroom like an apple tree an apple. The spawn is used to transfer mycelium on any material from which mushrooms grow such as substrate. Mother spawn is a mushroom fungus that grows on grain based medium. The production of mother spawn involves following steps:

- 10 kg of wheat or Sorghum grains are boiled in 15 litres water for 10-15 minutes and are allowed to remain soaked in hot water for 15-25 minutes without boiling.
- Water is drained off over wire netting to dry slightly.
- 120 g gypsum and 30 g lime are mixed with 9 kg of boiled grains.
- The gypsum prevents the sticking of grains together and lime adjusts the pH.
- The grains are filled into half litre milk or glucose bottles upto three-fourth capacity.
- Bottles are plugged with non-absorbent cotton and sterilized at 20-22 lb. psi (126° C) for 1 ½ to 2 hours.
- Sterilized bottles are taken out from the autoclave while still hot and are shaken to avoid clump formation.
- The bottles are immediately transferred to inoculating room or chamber and allowed to cool down overnight.
- Next day, bottles are inoculated with two bits of agar medium colonized with the mycelium of pure cultures raised either by tissue or spore by putting the culture bits just opposite to each other in the inner side of glass surface in the middle of the bottle.
- About 7-10 days after inoculation, bottles are shaken vigorously so that mycelial threads are broken and mixed with grains.
- Three weeks after incubation, the stock culture becomes ready for further multiplication of spawn.
- One bottle of stock culture is sufficient to multiply in 30-40 polypropylene bags or bottles. Inoculated bottles are incubated at 25 + 1° C.

Healthy and cleaned cereal grains

↓

Boil Grains in water (15-20 min.)

↓

Remove excess water on sieve

↓

Dry grains in shade (4 hrs.)

↓

Mix $CaCO_3$ (0.5%) and $CaSO_4$ (2%)

↓

Fill 300g grains in glucose/milk bottle

↓

Plug cotton and autoclave at 22 p.s.i. for 1.5 to 2 hrs.

↓

Inoculate growing mycelium of desired strain using laminar flow

↓

Incubate in BOD at 23 ± 2^0C for 20-25 days

↓

Master spawn is ready

3.7.2 Production of commercial spawn of different mushrooms

Commercial spawn can be prepared in polypropylene bags. Normally for half and one kg spawn the bags should be of 35 x 17.5cm and 40 x 20cm size, respectively. Polypropylene bags should have double sealing at the bottom and after filling the grains they are plugged with the help of a PP neck and non absorbent cotton. The bags are then sterilized at 22 lb p.s.i. pressure for 1.5 to 2 hours. Autoclaved bags are shaked well before inoculation so that the water droplets accumulated inside the bags is being absorbed by the grains.

The sterilized bags are kept on the laminar flow under U.V. tube for 20-30 minutes. Ten to fifteen g of grains from master spawn bottle is inoculated per bottle under aseptic condition or one bottle of master spawn is sufficient for inoculating 25 to 30 commercial spawn bags. Inoculated bags are again shaked so that the inoculum is well mixed with other grains. Then the bags are kept in incubation room for mycelium spread. During incubation the bags are regularly examined for mould infestation. Contaminated bags should be immediately removed before discarding the bags to avoid build-up of contamination in the vicinity. Normally it takes 15-20 days for complete spread of mycelium on the grains. Fully colonized bags should be kept in cold room (+40 C) for future use. The spawn of button mushroom, Pleurotus can be stored at this temperature. However, neither the culture nor spawn of Volvariella, Ganoderma and Calocybe is stored below 150 C

Unit 4

Development of Infrastructure for Button Mushroom

4.1 Low cost button mushroom house

Button mushrooms can be cultivated commercially anywhere as essential environmental conditions required are easy to maintain and requisite raw material, substrate and supplement are readily and locally available. Mushroom cultivation involves outdoor and indoor activities.

Low cost button mushroom house is normally made with bamboo frame and synthetic fiber cloth material, from inside and outside using paddy straw insulation in between, which tends to give good results under hill conditions for seasonal growing.

The floor of low cost button mushroom house is insulated with insulating material of certain thickness using thermacol or glass wool where insulation is protected by PVC sheet, below and above from moisture.

Grower can also construct low cost mushroom by building it in between two rows of bamboo trees with spacing of around 2.7 x 2.7 m. For mushroom shed construction, bamboo stems can be used to cover the shed. The side walls can be made from leaves, supported by bamboo stem frame. In this the roof can be made from stem and leaves.

Each shed can be portioned at the centre into two rooms of certain measurements with only single front door for each room which can be made from polythene sheet that can be fixed to stem frame. The space in each room should be such that multitier rack made of bamboo should accommodate minimum 50 mushroom bags. In this, the floor of the shed can be made with river sand upto height of 10 cm.

Cultivator can also construct semi permanent low cost button mushroom unit structure for button mushroom cultivation where simple rooms can be constructed using cement. The cropping rooms will have cemented floor, cemented walls, cemented or thatched materials ceiling where there could be arrangement for forced air circulation inside it. The room can be air tight to make air handling system work properly to obtain required air changes.

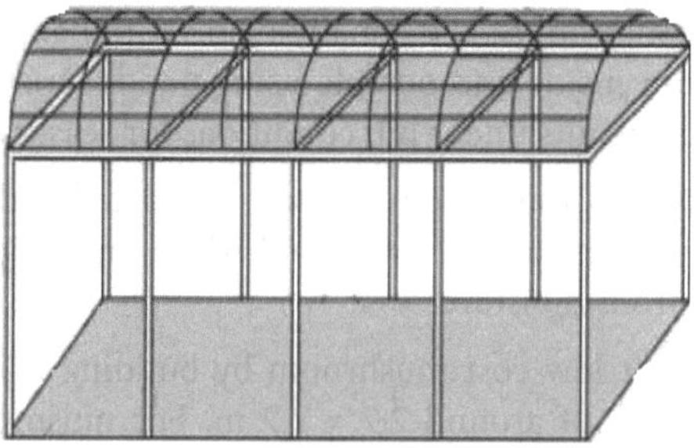

In low cost button mushroom cultivation room, grower can install heavy-duty coolers to lower the temperature in summers. Small growers can start mushroom cultivation in simple room/garage/store room with button mushroom cultivation.

4.2 Temporary/ Kaccha shed for button mushroom

Raw materials used for making temporary shed

- Bamboos
- Paddy straw
- Sutli
- Polythene

Grower can construct temporary shed for growing button mushroom by use of bamboo trees by maintaining required spacing of around 2.7 x 2.7 m. For cultivation of button mushroom, the kaccha shed can be constructed using bamboo stems, paddy straw, sutli and polythene.

The roof of kaccha shed can be made of bamboo, with paddy straw or PVC sheet. The side walls of the kaccha shed can be made from mud, wood, leaves, supported by bamboo stem frame. In this the roof can be made from stem and leaves.

Each shed can be portioned at the centre into two rooms of certain measurements with only single front door for each room which can be made from polythene sheet that can be fixed to frame. The space in each room should be such that rack made of bamboo/wood should accommodate minimum 50 mushroom bags. In this, the floor of the shed can be made with river sand upto height of 10 cm.

Kaccha structure should be constructed such that it can hold 16 tons of compost spread on its beds with depth of around 4-5 inches, so that required compost should yield around 50 kg of mushrooms daily in a season.

4.3 Permanent/ Puccka shed for button mushroom

Grower can grow button mushrooms in a permanent room with pukka shed, that will have following features:

Roofs

- The roof of permanent/pukka structure for growing mushrooms comprises of RCC of 12-15 cm thickness.
- The inside of the roof is cemented with plaster finish, for application of insulation on the walls.
- The roof from outside is protected by tarring it on top, which is followed by 10 cm thick loose soil, with 5 cm thick mud capping and tiles.
- The mud and tiles on roof top will protect roof from weathering effects of rain and ensures longer life of insulation, thereby preventing seepage of moisture in room in rainy season.
- In hilly areas where rainfall is heavy, roof should carry slanting GI sheet over the insulated RCC roof, to allow rain water to go down.

Walls

- The walls of permanent/pukka structure for growing mushrooms are made of brick of about 22.5 cm thickness, with cemented plaster.
- In this, insulation sheets are fixed on the walls with the use of hot coal tar.
- Here, holes are drilled on four corners of sheet/inside the cement wall for expansion fasteners which are fixed by screwing in nail with 4"-5" long steel wire tied on its head.
- The wire hangs out of the sheet to be used for tightening of wire net fixed on top of the insulation.
- The layer of cement plaster is applied 2 cm on top to give smooth finish.
- Bituminous paint is applied on cement plaster as vapour barrier.
- The painting can be avoided in cropping rooms if cook out is not done by steam.

Floor

- The floor of permanent/pukka structure for growing mushrooms should be well laid and strong enough to take heavy load of metal racks, that are kept inside for growing mushrooms.
- In this, floor should be insulated with insulating material of around 5 cm thickness using thermocol/glass wool.
- In this, insulation should be protected by PVC sheeting which can be below and above, to save from moisture.
- The insulating PNC sheet should be covered with wire mesh and concrete floor of around 5cm thickness should be laid on top, for good finishing.
- The floor should have slope towards the entry point for discharge of cleaning water and placement of formalin trough for foot wash.

- The trough is connected near the wall to an exhaust drain to carry washings from the room.
- In this, water discharge hole is protected to prevent from leakage of air from growing room.

Doors/vents

- The doors of permanent/pukka structure for growing mushrooms should be made of wood or angle iron frame covered on inside and outside with aluminum sheets/GI sheets with insulation of 5-7 cm in the centre.
- The doors will have rubber gasket lined on inner periphery so that door becomes air tight when closed.
- The door will operate on hinges, with strong locking latch for opening and closing.
- In this, exhaust vents are fitted with wire net, louvers and insulated lids, where louvers allow CO_2 laden air to exhaust under good pressure from blower.

4.4 Growing of button mushroom throughout the year

Market for mushroom is growing rapidly due to its nutritious values and special tastes. The main consumers of mushrooms are Chinese food restaurant, hotels, clubs and households. The growing domestic and export market and delicacy and food value provides extensive and good potential for cultivation of mushroom. There is also growing market for processed, dried and packed mushrooms as their shelf life is longer.

Cultivation of button mushrooms started in 16th century. However, on commercial scale, the cultivation was initiated in Europe around 17th Century. Many farms for production of button mushrooms were established and button mushroom variety still dominates the world production and consumption. India, with its diverse agro-climate conditions and abundance of agricultural wastes, has been producing mushrooms, mainly for domestic market, for more than four decades. Commercial production picked up in nineties and several hi-tech export oriented farms were set up with foreign technology collaborations.

Button Mushroom is most popular mushroom variety grown and consumed across the world. In India, its production earlier was limited to winter season, but with technology development, these are produced almost throughout the year in small, medium and large farms, adopting different levels of technology. The species being grown in most farms is white button mushroom which belongs to class Basidiomycetes and family Agaricaceae.

Growing white button mushrooms doesn't require sunlight, which is good for indoor growers whose windows are full up with plants. Such growers can grow these varieties at any time of year, with winter actually preferable, making for great gardening opportunity when everything outside is cold and bleak. Growing white button mushrooms takes spores, tiny microscopic things that will grow into mushrooms.

White button mushrooms grow best in nitrogen-rich manure, like horse manure. To create an indoor bed for these mushrooms, grower can fill a wooden box that's at least 6 inches deep with manure and leave a few inches of space below the rim of the box. Spread the inoculated material from the kit on the top of the soil and mist it thoroughly.

In button mushroom, vegetative mycelium is composed of many inter-woven sepatate hyphae. The reproductive phase in this is initiated by formation of small knob like swellings at different points of interwoven mycelial strands. Such swellings increase in size and break through the surface of substratum as small balls constituting the button stage. A matured basidiocarp is whitish in colour having thick short stipe with annulus. The stipe supports the pileus which appears as a hat like expansion. On underside of pileus, number of radiating gills or lamella are present which are pink when young but purple-brown when mature.

In India, button mushrooms are grown seasonally and in environment controlled cropping houses. White button mushroom requires 20-28^0 C for vegetative growth and 12-18^0 C for reproductive growth. Besides that it requires relative humidity of 80-90% and enough ventilation during cropping. Seasonally, it is grown during the winter months in the north-west plains of India and for 8-10 months in a year on the hills. However, with the advent of modern cultivation technology it is now possible to cultivate this mushroom anywhere in India.

The cultivation of white button mushrooms throughout the year under controlled condition is restricted to few commercial units and much of production is under natural conditions during the winters. Majority of growers in India do not have pasteurization facility and other sophisticated machinery/infrastructure for round the year production of white button mushroom. As such, button mushroom is cultivated seasonally when climatic conditions are favourable and production expenses are minimum.

In India, many growers in Haryana, especially in Sonepat, Ambala and Hisar have revolutionized the cultivation of white button mushroom by adopting very simple and cheap technology of construction of mushroom houses. Seasonal growing of white button mushroom in Haryana and Punjab has many advantages like nearness to market, easy and cheap availability of raw material coupled with utilization of family labour. The growers in HP do not use compost prepared by long method because pasteurized compost is readily available from mushroom projects located at Solan and Palampur.

Raw Materials Required:

1. Wheat/ rice bran
2. Poultry waste
3. Urea
4. Potash
5. Choker
6. Oil cakes
7. Molasses
8. Gypsum
9. Peat soil
10. Soil
11. Spawn
12. Waste paper/newspapers

Unit 5

Spawning and Casing

5.1 Transfer of Compost in composting room

Composting is necessary process to get selective and appropriate substrate for cultivation of mushrooms which belongs to genus Agaricus. Compost provides nutrients which is required for mushrooms to grow. There are two types of material that are generally used for mushroom compost, most used and least expensive being wheat straw-bedded horse manure.

In the process, materials are subjected to the natural action of many different microorganisms. During the process, soluble sugars are degraded and the temperature increases. It favors the development of thermophilic microorganisms, resulting into the generation of selective compost for the mushroom.

Synthetic compost is made from hay and crushed corncobs, although the term often refers to any mushroom compost where main ingredient is not horse manure. Both types of compost require the addition of nitrogen supplements and a conditioning agent, gypsum.

Compost can be prepared in Phase I and Phase II composting process where phase I compost results complete when mixture becomes dark brown with sweet smell, soft straws and raw ingredients are able to hold water with water moisture content at around 68%-74%. The phase II begins with pasteurization after 15-20 days of composting to kill bacteria and weed seeds and remove ammonia.

The compost after preparing will be transfer in special room, to have pasteurization of 8 hours with temperature around ± 60^0C and for conditioning period of 7 days at 45^0C temperature, to clear volatility of ammonia from air.

5.2 Different methods of spawning

5.2.1 Spawning

Mushroom compost can be inoculated with mushroom spawn for mushrooms to grow. In mushroom, fungus mycelium is white, thread-like plant often seen on rotting wood or moldy bread. Mycelium gets propagated vegetatively like separating daffodil bulbs and getting more daffodil plants. Mushroom spawn is mushroom mycelium growing on given substrate which serves as seed or planting material in mushroom cultivation. Spawning is inoculation of culture in substrate or compost, it is the actual planting of spawn and requires much care depending on the species of mushroom and the technology being followed.

Spawn makers start spawn-making process by sterilizing a mixture of rye grain plus water and chalk; wheat, millet, and other small grain may be substituted for rye. Sterilized horse manure formed in blocks was used as growth medium for spawn upto 1940, and this was called block or brick spawn, or manure spawn; such spawn is uncommon now. Once sterilized grain has a bit of mycelium added to it, the grain and mycelium is shaken 3 times at 4-day intervals over a 14-day period of active mycelial growth. Once the grain is colonized by the mycelium, the product is called spawn. Spawn can be refrigerated for a few months, so spawn is made in advance of a farmer's order for spawn.

Spawn is distributed on the compost and is thoroughly mixed in compost. For years, it was done by hand, broadcasting the spawn over the surface of compost and ruffling it in with small rake-like tool. In recent years, for bed system, spawn is mixed in compost by special spawning machine which mixes compost and spawn with tines or small finger-like devices. In a tray or batch system, spawn is mixed into the compost as it moves along a conveyer belt or while falling from a conveyor into a tray. The spawning rate is expressed as a unit or quart per so many square feet of bed surface; 1 unit per 10 ft is desirable. The rate is sometimes expressed on the basis of spawn weight versus compost weight; a 2 percent spawning rate is desirable.

Once the spawn has been mixed throughout the compost and the compost worked so the surface is level, the compost temperature is maintained at 75°F and the relative humidity is kept high to minimize drying of the compost surface or the spawn. Under these conditions the spawn will grow – producing a thread-like network of mycelium throughout the compost. The mycelium grows in all directions from a spawn grain, and eventually the mycelium from the different spawn grains fuse together, making a spawned bed of compost one biological entity. The spawn appears as a white to blue-white mass throughout the compost after fusion has occurred. As the spawn grows it generates heat, and if the compost temperature increases to above 80° to 85°F, depending on the cultivar, the heat may kill or damage the mycelium and eliminate the possibility of maximum crop productivity and/or mushroom quality. At temperatures below 74°F, spawn growth is slowed and the time interval between spawning and harvesting is extended.

The time needed for spawn to colonize the compost depends on the spawning rate and its distribution, the compost moisture and temperature, and the nature or quality of the compost. A complete spawn run usually requires 14 to 21 days. Once the compost is fully grown with spawn, the next step in production is at hand.

5.2.2 Method of Spawning:-

Spawning is inoculation of culture in substrate or compost. It is planting of spawn and requires much care depending on the species of mushroom and the technology being applied. However, there are several methods of introduction of mushroom to substrate such as:

Double layer spawning

- ❖ In this, spawning is done by scattering spawn on tray beds once when half-filled with compost after complete filling of tray.

- ❖ The spawn is gently pressed down with forefinger uniformly each time and trays are covered with newspaper sheets.

Top spawning

- ❖ Once the tray is filled up to brim with compost, spawn is planted above the surface and a thin layer of compost is spread over as if spawn is left at top of compost which dries up quickly.
- ❖ If the top will not dry out it is not necessary to plant the spawn deep.
- ❖ If compost is moist enough it is advisable, not to go deep.

Thorough spawning

- ❖ In this, whole of spawn grain or shredded manure spawns are mixed throughout the compost.

Shake-up spawning

- ❖ A few workers have reported better yield by shake-up spawning.
- ❖ After a week of spawning, compost is thoroughly shaken up, and replaced in shelves or in trays.
- ❖ After that either it is cased at once or a few days later.

Active mycelium spawning

- ❖ In this method, fully run trays of spawned compost are used for spawning further trays.
- ❖ In this, one such tray is used for spawning of many trays.
- ❖ In this method, there are more chances of contamination.

Spot spawning

- ❖ In this method, grain spawn can be kept in holes at certain distance with pointed stick or fingers.
- ❖ In this, the distance between the hole is immaterial and how the hole is made.
- ❖ In this, care should be taken by growerr to ensure close contact of inoculum with surrounding compost to ensure fast development of mycelial strands.
- ❖ Here, cavity is covered with the compost.

Super spawning

- ❖ In this, fresh pasteurized trays are planted as required.
- ❖ Such method is known as preliminary spawning.
- ❖ After two weeks, compost from each tray is mixed with compost from newly pasteurized trays which gets packed firmly, so it is known as super spawning.

From the above methods, grower can use any method, but double layer and thorough spawning methods should be preferred for spawning in low temperature. Spot spawning method can be used for spawning with compost spawn. Once the compost is spawn, compost surface is covered with old newspaper sheets which are sprinkled with water to provide humidity as no water is directly added to compost during spawn run.

5.3 Casing material

5.3.1 Casing

Casing is top-dressing which is applied to spawn-run compost on which mushrooms form. Clay-loam field soil which is a mixture of peat moss with ground limestone, or reclaimed weathered, spent compost can be used as casing. Casing does not need nutrients as it act as water reservoir and place where rhizomorphs form. Rhizomorphs look like thick strings and form when the very fine mycelium fuses together. Mushroom initials, primordia, or pins form on the rhizomorphs, so without rhizomorphs there will be no mushrooms. Casing to be pasteurized to eliminate any insects and pathogens it carries. It is noted that casing should be distributed so that depth is uniform over the surface of compost as it allows spawn to move in and through the casing at same rate and allow mushrooms to develop at the same time. It should be able to hold moisture since moisture which is essential for development of firm mushroom.

Managing the crop after casing requires that compost temperature be kept at around 75°F for up to 5 days after casing, and relative humidity should be high. After this, compost temperature should be lowered at about 2°F each day till small mushroom initials form. Throughout the period following casing, water must be applied intermittently to raise the moisture level to field capacity before mushroom pins form. Knowing when, how, and how much water to apply to casing is an art form which readily separates experienced growers from beginners.

5.3.2 Types of Casing material

Introduction

In ancient times, selection of casing materials depend on availability of different materials in particular area. In India, top soil from cultivated field was tried as casing material, but later on clay, loam, sand and mixture of ash, stone, gravel and spent compost were tried.

Various materials have been tested as casing substrate across the world by different workers. Bewley compared the suitability of variety of materials as casing and found an increased yield of mushrooms by addition of peat soil.

Casing Material Observations

- The use of moss and mixture of garden soil and peat has been reported and was found that non-fibrous peat gave better yield.
- The vermiculite, sand, peat and local soil successfully act as casing materials and as a result, sphagnum peat became standard casing substrate.
- As reported, mixture of sand and peat in equal volumes produce better and good quality fruit bodies.
- As compared, hygromull, styromull and polyethers as replacement for peat dust found hygromull at 25% concentration which improves mushroom yield.

- It was compared that pine bark and peat (1: 1) get comparable yield in such media, and also the chances of phytotoxicity of soil with bark.
- The suitabiUty of sphagnum moss growing abundantly in hill areas results that mixture of sphagnum moss + loam soil in ratio of 2:1 gave best results.
- The best yields of mushrooms is given by using topogenous peat as casing material, but best results obtained with regards to mushroom yield with mixture of peat + river mud in 2:1 ratio shows more mushroom weight with strong decomposed peat in form of casing as compared to decomposed peat.
- It was adviced that the use of spent compost for purpose of casing is good.
- A casing material was developed based on a year old spent compost, sand and soil in ratio of 4:1:1 after treating it with steam, nemagon or formalin.
- The number of irregularities during crop production leading to poor yield and appearance of crop in patches with the use of year old spent compost for reuse of spent compost as casing material.
- The paper mulch and two years old spent compost, gives maximum yield in certain conditions.
- There will be an increase in mushroom yield with ageing of spent compost.
- The highest mushroom yield was obtained with combination of spent compost + 8% calcium carbonate.
- Spent mushroom substrate could be recomposted after cropping of mushrooms to produce peat like material and can be modified by leaching to be used as casing material, provided it should be free of potentially harmful organisms and met the required physical and chemical attributes of casing.
- The chelated spent mushroom substrate as casing material for production of A. bisporus which found casing prepared from spent mushroom substrates treated with EDTA and blended with peat was most productive.
- The use of weathered spent compost for casing was practically economical material if handled properly.

Casing Layer

Casing layer is a layer of moist material which can be either organic, inorganic or combination of both, which can be kept on top of colonized substrate prior to fruiting. Its main function is to aid with moisture retention in substrate. Some casings also contain biological components, microorganisms that are necessary for fruit body formation in some species.

Many different materials can be used for casing layer which includes:

- peat moss
- garden soil
- sphagnum moss
- coco coir

Peat moss

- ❖ It is commonly used casing material which is cheap and widely available.
- ❖ It is excellent at retaining moisture.
- ❖ It is naturally acidic, so growers often add calcium carbonate to the mixture in order to buffer the pH which helps to mitigate the chances of contaminates flourishing in moist casing layer.

In this, a ratio of calcium carbonate is around 2-3% by wet weight of casing layer.

Vermiculite

- ❖ It is also a common casing material.
- ❖ It is an inorganic material which is added to soil to help potted plants retain moisture.
- ❖ It is very resistant to contamination and can be added directly to top of substrate without any consideration for sterilization or pasteurization.

Many home growers uses hydrated lime (calcium hydroxide) as a way to buffer the casing layer.

5.4 Care and maintenance of mushroom house

5.4.1 Introduction

- ❖ Mushroom home environment is favorable to growth of mold fungi, so care has been taken o prevent and keep the surroundings hygenic.
- ❖ In addition, temperature and humidity are critical factors that must be maintained to ensure that the fungus can produce a good production.

a) Environment

Mushroom grown in rooms need clean, orderly and free of bacteria, animal and insect pests, so mushroom house should be treated with care and proper.

The controls that are carried out to maintain the cleanliness in mushroom house includes:

Putting chalk on shelf/surroundings

- ❖ Lime should be sown on the empty shelves before putting new bag of mushrooms.
- ❖ Applying lime or chalk on shelves and surroundings will intend to kill the bacteria that are found in the vicinity of shelves and around mushroom house surroundings.
- ❖ The area around the bag rack infected mushrooms will also be placed with chalk bag to avoid infection of other fungi.

Placing citronella in mushroom house

- ❖ Citronella can be kept in the mold which is used to drive mushroom home that kills small insects and other pests found in the vicinity of mushroom houses.
- ❖ Citronella will be mixed with water and placed in a bottle, drilled with a hole, so that smell of citronella can be spread around the whole mushroom house.
- ❖ Citronella bottles will be suspended at the front of each pole shelf mushroom bags.

Mushroom bags are damaged and expired useful life

- ❖ Bag of mushrooms that span expired, damaged, infected, and problematic to be separated and taken out of bag for new media can be brought in mushroom house to replace old media bag.
- ❖ Mushroom bags also prevent bacterial infection from a bag that had been infected.

Provision of suitable space

- ❖ Space in mushroom house should be sufficient for staff/workers to control all activities that occur in mushroom homes related to storage bags, harvesting of mushrooms, water treatment and in managing media bags whose span has expired.

Mushroom Shelves

- ❖ In mushroom house, mushroom shelves built and height must have suitable area for storing luggage mushrooms.
- ❖ In this, the racks should be placed at the right distance and convenient to control the activities of the new media storage bag, to pick mushrooms cooked and spray water.
- ❖ In case of small-scale planting, distance between the shelves should be such so as to accommodate employee/worker to bring a basket and to allow plant maintenance activities.
- ❖ In case of large-scale planting, distance between the rack should be such that it allows harvest work done by using wheelbarrow by the workers.

b) Temperature and humidity in mushroom house

- ❖ Temperature and humidity are important aspects to be considered in controlling the mushroom house as it is affects the growth of mold and mold that would be obtained.
- ❖ Natural environment during the hot/rainy day which affects temperature and moisture directly in mushroom house should be maintained.
- ❖ The selection of design and materials used for mushroom houses are also factors which affects the temperature and humidity in mushroom house.
- ❖ Such factors are important as it will affect the growth of fungi and media itself in resistant bag.

Concrete Buildings

- ❖ Factors to be affected by humidity and temperature of heat trapped inside the mushroom house.
- ❖ As per the convection process, hot air has low density, so it will rise to the top while cold air in air with high density decreases.

- The bag of media will produce heat by splitting spores which reduces moisture in mushroom house.

Temperature and humidity are conducive to mold growth

- Temperature suitable for mushroom house is at room temperature between 25-27'c.
- In mushroom house, appropriate humidity is between 80-83 percent humidity.
- Measuring devices should be placed in temperature and humidity in mushroom house to find out the actual temperature and relative humidity.
- The measuring device should be placed at an angle of central mushroom house.

Method of temperature and humidity control

Method of temperature control is done by several ways such as:

- Water Pipes Shot
- Artificial Rainwater
- Fan Air
- Spray water pipe

Thus, the first rule of water spray is used to solve this problem. Shot will be done by watering at the end of the media bag and mushroom house floor. This method will reduce the amount of hot air that rise to the top.

Water spray hose will do as much as 5 times in a day, at about 11.00am, 3.00 pm, 5.30pm and 7.00am 12.00PM and on a hot day with the normal temperature 28 ° -32 ° to maintain the temperature at 25 ° -30 °.

On rainy days during the day, a lecture on mushrooms and flowers bag will be reduced, only the floor will be given more water.

- Techniques of water spraying the right:
- Standing in the middle path between the rack holding the hose mushrooms.
- Spraying the floor to really wet and damp.
- Alternate spray-hose spray into the bag of mushrooms to moisten a little bag of mushrooms.
- Avoid spraying too much water in the mold to prevent interest rates from the fresh mushrooms, grow faster and chap.
- On hot days, mushroom house should be watered more and on a rainy day, the water is sprayed into the bag of mushrooms should be reduced.

Fan

The fan is used to bring in cool air from the hot air out and cool in the mushroom house. The process of convection occurs due to turbulent air from the fan. Some of the cool air will come down, while the hot air will escape through the hole in the roof.

Artificial Rainwater

- ❖ Artificial rain will be installed during hot weather because of the radiation emitted by the sun as it will warm the air inside the mushroom house.
- ❖ The water will reduce the radiation to penetrate the roof.
- ❖ In such case, the hot air entering the fan will be issued.
- ❖ The water will reduce the radiation to penetrate the roof and cool the roof, thereby reducing heat in the mushroom house.
- ❖ Water to be sprayed as per the temperature and humidity in mushroom house that is constantly changing as per change in environment.

Unit 6

Biotic and Abiotic Stresses

6.1 Symptomatology and Management of fungal diseases

6.1.1 Introduction

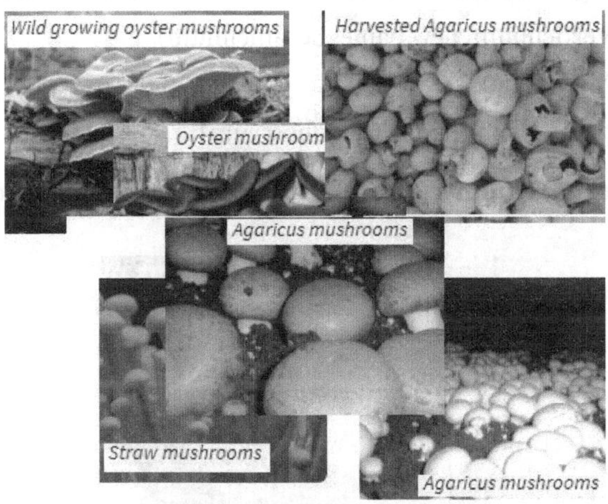

Mushrooms are fleshy fruiting bodies of fungi which include edible species in genus Agaricus, Pleurotus, and Volvariella. They are highly variable in appearance as per their stage of development and variety. Agaricus mushrooms have cap and stem, where cap is a pale grey or brown in color and is round while immature, but flattens out as it reaches maturity and can reach a diameter of 5–10 cm. Pleurotus mushrooms may not have a stem and instead may be laterally attached to a growing substrate such as the bark of a a tree. Pleurotus mushrooms are smooth and elongated and can reach 4–15 cm in diameter. Volvariella mushrooms are small, with pink gills and a characteristic sac-like covering at the base of stem. The cap can reach 5–15 cm in diameter.

Factors that cause diseases in mushrooms:

Biotic factors:

❖ Fungi

❖ Bacteria

- Viruses

Abiotic factors:

- Air
- Temperature
- Nutrition

6.1.2 Fungal Diseases

Fungi were earlier in Plant Kingdom and then closely related to animals and plants. Fungi feed by absorbing nutrients from organic material on which they live. Fungi digest their food before they absorb it by secreting acids and enzymes. They are essential part of carbon cycle and are responsible for breaking down dead organic matter which allows nutrients to cycle through the ecosystem. Many plants cannot grow without fungi that inhabit their roots and supply them with essential nutrients.

There are many fungal pathogens, but only few of them causes severe crop losses such as:

- Dry Bubble (Verticillium, new name: Lecanicillium)
- Wet Bubble (Mycogone)
- Cobweb (Dactylium, new name: Cladobotryum)

Disease: Dactylium; Hypomyces rosellus (Cladobotryum dendroides, Daktylium dendroides)

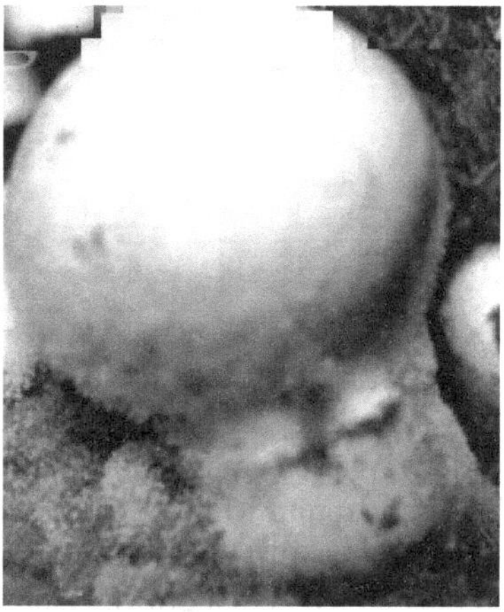

Symptoms:

- Webbed cottony mycelial growth on surface of casing and mushrooms
- mycelium is usually white but may be gray or pink in color
- infected mushrooms will develop soft and watery rot

Information:

- The disease result in cobweb mold as of the outward appearance of mycelial growth, which looks as white cobweb.

- ❖ The cobweb mold appears on casing layer as small and white spots grows very fast all over the casing surface, colonizing all mushrooms on its way.
- ❖ The diseased mushrooms become covered with white and fluffy mold mycelium, weaken, become brown and rot.
- ❖ In this, mycelium threads of mold gradually thicken and become think layer of mold.
- ❖ The colour of mycelium threads becomes reddish-purple which changes to yellow on later stages of development.
- ❖ During spore-forming, spore dust is easily spread by wind or watering the beds.
- ❖ In this, dactylium is easily spread by pieces of mycelium, which stick to personnel's clothes and implements.
- ❖ In this, growth of mold is restricted to casing layer and disease can spread very quickly if left untreated.

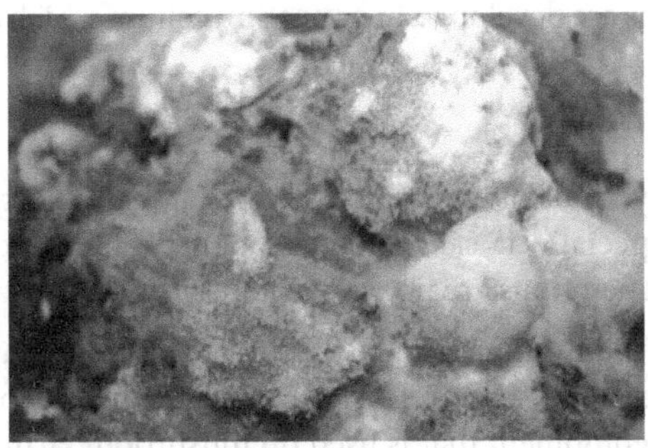

Precautions:

- ❖ The measures should be taken in order to prevent the occurrence of infection, which covers:
- ❖ The casing layer has to be prepared in a clean, disinfected room. If necessary, the casing layer can be watered with 2% of formalin solution or fungicide solution, right after it's been applied.
- ❖ The relative humidity and air temperature have to be maintained.
- ❖ The transfer of infection to the healthy mushrooms has to be avoided, by organized movement of pickers.
- ❖ Abide by the rules of sanitation and hygiene on the whole farm.

Management:

- ❖ Control of disease is done by good sanitation practices and cultural control methods.
- ❖ The casing should be kept clean and sanitized.
- ❖ All tools and equipment should be cleaned and sanitized before use.
- ❖ Keep beds free of any mushroom debris.
- ❖ Pathogen in susceptible to low humidity and can be inhibited by lowering the humidity in growing room

Disease: Green mold Trichoderma spp.; Penicillium cyclopium; Aspergillus spp.

Symptoms:

- A dense layer of mycelium is present on casing surface which is green in coolur.
- The mushrooms are developed in or near mycelium which are brown and gets cracked or distorted

Information:

- There are many kinds of green molds which mushroom grower come across during cultivation, especially if sanitation is not maintained on farm/room, and when technological operations are not carried out properly.
- The spores of these fungi are widely spread in casing layer and organic matter in various environments and are carried by wind, insects or mites, personnel on any inventory and equipment that is used on mushroom farm.
- Rats and mice that feed on mycelium on the surface of compost also carry such disease.
- Green fungi grows on wooden equipment, in compost, in casing layer, and on poorly prepared mycelium grains.
- It is one of the most common and destructive diseases of cultivated mushrooms.

Management:

- Prevention of disease is best achieved by good sanitation practices.
- Ensuring compost is properly pasteurized prior to use and sterilize any supplements

Disease: Verticillium spot Verticillium fungicola

Symptoms:

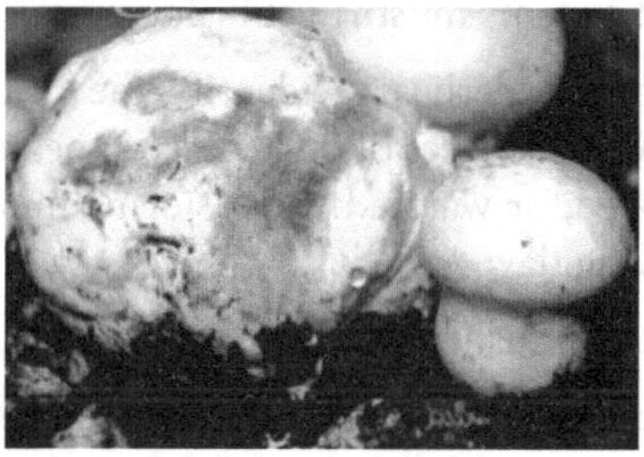

- ❖ Small spotting on mushrooms
- ❖ mushrooms deformed
- ❖ surface of mushroom becomes gray and fuzzy
- ❖ develops localized dead areas on cap
- ❖ pinched areas on cap
- ❖ fruiting body may no longer be recognisable as mushroom
- ❖ severe infection can cause deformation known as dry bubble which causes sporocarp to appear as large puffball like masses
- ❖ bubbles will become covered with gray
- ❖ fuzzy growth

Information:

- ❖ Verticillium disease is a mushroom fungus disease which often occurs and causes considerable damage to mushroom culture.
- ❖ Verticillium that has appeared on an early stage of mushroom development and wasn't treated in any way, can colonize and practically destroy the whole yield. The disease symptoms can be different, as it depends on the time the infection occurred.
- ❖ The disease may be confused with bacterial infections as it spread by flies and/or infected equipment

Management:

- ❖ Bubbles should not be disturbed by touching and can be destroyed using salt.
- ❖ Salt should be placed in plastic cup and then kept over bubbles to dry them out by providing barrier to prevent spores spreading.
- ❖ Fungicides are available for treatment of Verticullium diseases of mushrooms but many have toxic effects on the mushrooms as they themselves are fungus

Control Measures:

The control measures of such disease include:

- ❖ maintaining the hygiene on the farm;
- ❖ preventing the appearance of flies and mites on the farm;

- the preparation and storage of the casing mixture in a clean room. Mushroom wastes, outside soil, insects and rodents mustn't get inside;
- Teaching the pickers and orderlies to identify the disease and treatment of the infected mushrooms. When sources of the disease appear, they must be strewed with salt and carefully removed, or watered with a formalin solution, strewed with lime and removed the following day;
- The removal of all infected mushrooms before harvesting or watering;
- Processing of the infected batches as soon as possible by steaming, or using a strong disinfectant;
- Immediate removal of all wastes from the farm.

Disease: Wet Bubble; Mycogone perniciosa

Symptoms:

- It shows wet bubble at different stages of mushroom development.
- The infected sporophores and sclerodermoid masses occurs at different stages in development of sporophores.
- The infection before differentiation of stipe and pileus results in selerodermoid form, while infection after differentiation results in thick stipe with deformation of gills.
- The mushroom shows white mouldy growth, leading to putrifaction with golden brown liquid exudate.
- Symptoms of infected sporophores results in tumorous form or malformation infection at later stages.
- It shows 1/3rd infection at base of stipe in apparent healthy sporophores in form of black streaks.
- The pathogen has short, curly, pure white fluffy mouldy growth on malformed mushrooms.
- The deformed sporophores show black circular area below the upper layer.

Information:

- ❖ Wet bubble in white button mushroom incited by Mycogone perniciosa Magn. is one of serious diseases from almost all major mushroom growing areas across the world.
- ❖ Bubbles or mole, is responsible for heaviest losses in mushroom beds in.
- ❖ The disease also reported to assume serious proportions in other main mushroom growing areas

6.2 Symptomatology and Management of bacterial and viral diseases

6.2.1 Bacterial diseases in Mushroom

In mushrooms, bacterial diseases can be:

- ❖ Bacterial blotch: caused by Pseudomonas tolaasii
- ❖ Mummy disease: caused by Pseudomonas sp. Schisler
- ❖ Soft rot
- ❖ Bacterial rot
- ❖ Yellow blotch

Bacterial Blotch Disease

Information:

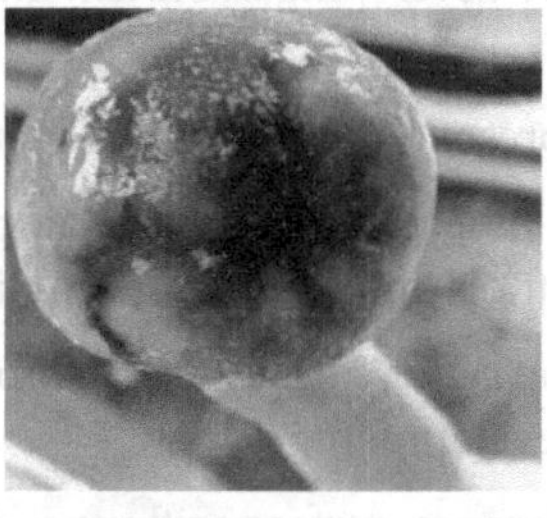

- ❖ The bacterial disease appears on fruit bodies of A. bisporus, A. bitorquis, Pleurotus species, Volvarella species, Lentinus edodes, Flammulina velutipes and Auricularia species.

- The bacterial pathogens induced many symptoms like blotch, mummy, pit, drippy gill, soft rot, yellowing and immature browning, but bacterial diseases reports only on fruit bodies of A. bisporus and species of Pleurotus and Auricularia.

- Bacterial blotch may be endemic on mushroom farms, which damages mushroom quality and posing potential yield loss.
- Presence of disease is probable if surfaces of mushrooms do not dry following watering, irrespective of season.
- Blotches appear when the mushrooms are in early button stage, but can appear on mushrooms at any age.
- If moisture conditions favor the disease, spots enlarges and coalesce, which covers entire mushroom cap.

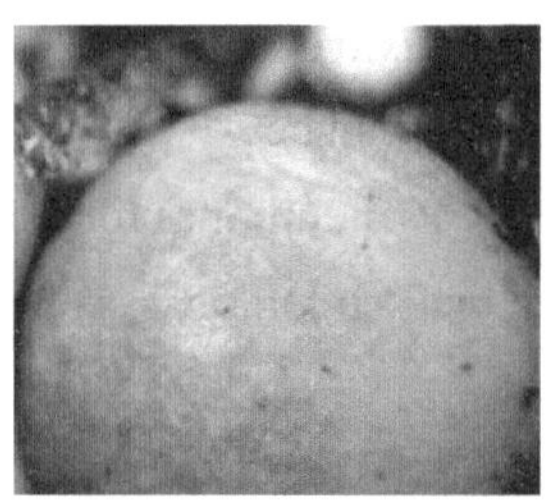

- In this, spotting is observed at or near the edge of mushroom caps at contact points between two mushroom caps, at crevices in clusters of mushrooms, or when mushroom caps remain wet for long hours.
- If dry conditions occur after developing of blotch, infected caps may crack radially as the mushroom expands.
- The bacterial pathogen is present in many casing material, even after pasteurization.
- Once the disease occurs, blotch-causing bacteria gets spread by splash-dispersal during watering, on using tools and by mushroom flies and nematodes.
- Bacterial blotch develops on outer surface of mushroom, on cap or stem at any stage of mushroom growth or development.
- Bacteria splashed on mushroom surface will reproduce in moist conditions, when water condenses or remains on mushroom surface for number of hours.

- In this, cap surface is cooler than surrounding air when water transpires from mushroom due to active mushroom growth and transpiration produces cool surface.

Common Names

- Bacterial blotch, Brown blotch, Bacterial spot

Occurrence:

- Occurrence of disease is linked with the size of bacterial population on mushroom cap, rather than on population in casing.
- The bacterial blotch of mushroom fruit bodies is caused by bacteria called Pseudomonas tolaasii and by species of Pseudomonas bacteria genus.
- The spread of bacteria start from casing layer but when sanitary regulations are not observed, they colonize the compost after pasteurization.
- Such disease occurs in summer-autumn time, when humidity and temperature of air is quiet high.
- Due to improper conditions of cultivation, insufficient ventilation after irrigation causes mushrooms to remain moist for longer period, which promotes development of disease.
- In favorable conditions, bacteria reproduce very fast on moist mushrooms and with vital activity, mushroom caps become covered with yellow-brown spots.

Symptoms:

- Circular but irregular, yellowish spots on or near margin of cap of wet mushroom which expands under high humidity conditions and coalesce to form big rich chocolate brown spots.

Management

- Manipulation of relative humidity, temperature, air velocity and air movements are of great significance in managing such disease.
- For this, the temperature should be above 20°C and relative humidity should not be more than 85%.
- Ventilation and air circulation after watering ensures quick drying of mushrooms.

- Temperature fluctuations at higher relative humidity leading of water condensation to be avoided.
- Isolation of P. fluorescnes and other antagonistic bacteria results in 30%- 60% control of the disease.
- Selective bacteriophages results effective against P. tolaasii without any significant effect of P. fluorescens.
- Spraying of casing soil with mixture of P. fluorescens and bacteriophage results in more 80% control of blotch disease symptoms.
- Application of terramycin 9 mg per square foot, streptomycin, oxytetracycline, kasugamycin and kanamycin results effective in managing the disease.
- Pasteurization of casing soils by steam/air mixture and short wave length irradiation is effective in removing bacterial pathogen, but over-heating should not be done, as successive invasion of moulds will results as very high.
- Use of water retentive acrylic polymers for casing soil mixture also reduces such disease.

Yellow Blotch:

Information:

- Bacterial blotch of Agaricus bisporus is caused by Pseudomonas tolaasii (brown blotch) or Pseudomonas gingeri(ginger blotch).
- Genus Pseudomonas(sensu stricto) has taxon of metabolically versatile organisms that are ubiquitous in soil and water which plays important role as plant, animal and human pathogens.
- The yellow blotch disease of Pleurotus spp. is caused by Pseudomonas agarici.
- The disease incidence is more under warm and humid conditions.
- The disease appears as blotches of varying sizes on pilei sometimes depressed, yellow, hazel-brown, fawn or orange in colour.

Symptoms

- Small, irregular, yellow spots appear on mushroom caps which expands and become dark-brown blotches on button mushroom and mature caps.
- Disease appears as blotches of various sizes in pilei, yellow hazel brown or organic in colour. The infected fruit bodies turn yellow and remain stunted, turn slimy and start giving foul smell.
- With late flushes, distortion of emerging caps is observed.
- Bacteria growing on cap surfaces cause affected tissue to have slippery feel.
- Stalks can be attacked.

Management:

- ❖ High relative humidity and continuous persistence of water film on surface of pilei, improves bacterial multiplication.
- ❖ Proper ventilation, careful watering and monitoring of temperature in mushroom unit, helps in limiting the disease incidence.
- ❖ Regular use of chlorinated water with 100-150 ppm of chlorine at 3-5 days interval helps in minimizing losses due to bacterial pathogen.
- ❖ The use of bacteriophages as control agent for plant diseases by bacterial pathogens are used in mushroom industry.

6.2.2 Viral diseases in Mushroom

Information:

- ❖ During recent years, viruses have increasingly been found along with fungi, where fungus acts as vector of the virus and also it acts as host of the virus.
- ❖ The virus disease is dangerous and are difficult to control.
- ❖ Discovering and finding infection of mushrooms with virus is hard and impossible as symptoms of virus aren't determinant and are very similar to symptoms of other diseases.
- ❖ The sign of symptom result from virus activity, where a sharp decrease in yield or complete absence of mushrooms is found, once hard work is done.
- ❖ In spawn run period, presence of virus is shown by diagnostic tests with use of microscope.
- ❖ The manifestation of other symptoms distinguishes virus infection from other diseases, which depends on combination of factors like; time of infection, amount of introduced diseased spores or infected mycelium, concentration and combination of kinds of virus particles present.
- ❖ Infection of mushroom culture with virus be identified by symptoms such as mycelium hardly grows in casing layer, and disappears or doesn't grow.
- ❖ Fruit bodies of very low quality appear late and grow in thick clumps in areas of the casing soil that haven't grown with mycelium;
- ❖ The mushrooms look like drum sticks, where very small caps grey or dirty cream in color have longer and deformed stipes, where fruit bodies are loosely attached to casing layer;
- ❖ Mushrooms infected with virus develop slowly, but their caps open quickly, releasing millions of spores infected with virus.
- ❖ The danger of virus infection results in loss of yield of single batch.
- ❖ Using diseased spores and remains of infected mycelium, virus spreads through whole farm and enters new cultures, that leads to suspension of production.

- Spores are easily carried by wind, insets, on implement, clothes and hands of personnel, hence the possibility for spores infected by virus moving from old rooms to new batches is very high.
- The size of damage depends on stage on which the infection occurred, so when virus enters mycelium during application of casing soil or later, then yield loss will be less and brownish mushrooms with deformation typical for virus disease are found at first break.
- In subsequent breaks, disease occur as bigger or smaller lesion focus, despite of sufficient mushroom yield, where batches present themselves as source of disease for other growing rooms.
- Infection which occurs before application of casing soil, has grave consequences, especially if spores or mycelium that contains virus fall in different places at same time.
- When mushroom mycelium is infected with virus during spawning, it results great yield loss.

Symptoms of Mushroom infected with virus

The general symptoms observed by mushroom grower on mushroom infected with virus will be:

1. Mycelium will not or hardly permeates casing layer and disappear after normal spread. In this case, mushrooms will appear only in dense clusters and gets mature very early.

2. Mycelium isolate from diseased sporophores on agar and shows slow and degenerated growth as compared with healthy mycelium.

3. Delayed appearance of pinheads of first flush is an important indication of disease and formation of fruit primordia below surface of casing layer. As these mushrooms appear above casing soil, pilei gets opened.

4. Symptoms of sporophores are highly variable.

There are certain abnormalilties that are found in mushroom such as:

- Slow mycelial growth, development of abnormal mushrooms.
- Slow development of pinheads, dwarfing.
- Delay in appearance of sporophores, reduced yield.
- Off-white colour of the cap and early maturity.
- Sporophores with elongated stems and small caps.
- Elongated slightly bent stipes, sometimes with small early maturing pileus.
- Premature opening of veil.
- Mushrooms are loosely attached to the substrate and at the slightest touch are pushed over.

- Accelerated post-harvested deterioration.
- Watery stipes, streaking in the stipes
- Stipes are spongy and quickly turn brown on cutting and show an abnormal structure.
- Thickened barrel-shaped stipes; the veil is attached to the thickest part of the stipe, thus lower than usually. Pilei are small and fat.
- Brown, slimy, cap occur owing to a secondary bacterial rot, stipes are sometimes tapering downwards during the first flush, sometimes a few light brown caps can be observed.
- Veils abnormal or absent, hard gills are common.

5) Particular musty smell perceives in growing room gets infested with disease. In case of Pleurotus, virus infection causes dwarfing or elogation of stipe, but no distinct symptoms are seen in Volvariella.

Management

Mushroom grower must make sure that they:

- harvest mushrooms in proper time, not allowing them to open;
- use filtered air during the pasteurization process, spawning and spawn run;
- use filters in order to keep spores in all growing rooms, when the farm is infected by virus;
- change filters after each batch, and steam them after treating the growing room;
- as most critical moments for infection takes place during unloading of pasteurization tunnels, spawning and filling of growing rooms, so such operations to be done after the equipment, machinery and room have been cleaned and disinfected;
- after spawning, it's best to cover the beds with paper, should sprinkle 2% formalin solution every 3-4 days, in order to kill all settling spores;
- prepare the casing soil, use only clean supplements, that are free of pathogenic organisms;
- apply casing layer using disinfected equipment;
- should disinfect the equipment after each bed is loosened, while loosening casing soil;
- use disposable crates or containers when harvesting mushrooms;
- thoroughly cleaned and disinfected crate before use, as one crate for many time use will increase the risk of infection and spreading of virus;
- steam growing room at compost temperature of 70°C for 12 hours, after the end of growing cycle;
- after steaming and immediate removal of spent compost from farm's area, should thoroughly clean and disinfect growing room and all equipment;

6.3 Identifications of insects and their managements

6.3.1 Introduction:

Mushrooms are the best cash crops for small farms and growing operations to invest as they develop their business. Since these edible fungi don't need sunlight, they're grown inside and in a carefully controlled environment. Apart from guaranteeing robust growth of fungi, these operations can be ideal breeding ground for number of pests that can decimate mushroom crop. Controlling and eradicating these pests should be main concern for mushroom growers.

6.3.2 Pest Identification:

The adult pests are small, delicate, greyish black, two winged insects which measures around 2.5- 3.0 mm in length with conspicuous bead-like antennae. Female pests have pointed abdomen that is frequently swollen with eggs, while males have prominent claspers at end of their abdomen. In these, larvae are dirty white with black head and are 6.5 mm long.

6.3.3 Pest Damage:

The larvae feeds on mushroom which enters from the stem end and move towards the cap portion. The most serious injury is caused which develops pin-heads, that turns yellowish brown which are not develop in normal mushrooms. Here large numbers of larvae feeds in compost during spawn run which also inhibit fruit body production by destruction of compost and mycelium. The incidence of pest starts in first week of February and reaches its peak in second week of March, with an overall average infestation of 40 per cent.

6.3.4 Types of Pests

In mushroom production, the main pests are flies, mites, and nematodes while other pests are mice, spring tails, thrips, beetles, and sowbugs. There are different types of pests that are involved in indoor operations, which damages different mushroom crops, so mushroom grower can easily identify such pests which includes:

- Cecid Fly – Under optimal conditions, this fly will multiply rapidly. In this, the adult cecid fly is rarely seen due to larvae handle reproduction. As legless maggots, the larvae are white or orange and spread throughout a mushroom farm by sticking to tools and equipment. They feed on the mycelium, stipe and gills of mature mushrooms.
- Phorid Fly – As creamy-white maggots, these insects will feed on mycelium. In early stage, they only eat selectively and are usually tolerable, but once they transform into fly, they are capable of transmitting fungal and bacterial diseases by the crop. Adult fly have hunchbacked appearance and are stocky looking and are highly active even while on land, often hopping erratically.
- Sciarid Fly – The most pervasive pest for many growers, sciarid fly larvae are voracious eaters that devour mycelium and compost. After about 15 days of feeding, they enter a pupae stage by spinning a web around themselves. Adults are most evident around light sources and have a black body, long antennae and gray wings. Infestations of sciarid flies, which also are called fungus gnats, manure flies or mushroom flies, cause mushrooms to become brown and leathery.

- Nematodes – They are normally considered as beneficial soil organism, which actively devour the fungi that are present in mushroom farms. These microscopic creatures are detected by degeneration of mushroom mycelium. The presence of nematodes shows problems in sanitation and pasteurization processes.
- Other Flies – There are number of other species of flies which are also an evident in mushroom growing operation.

1. Fly Pest:

Commercial mushroom fly pests include three dipteran families: Sciaridae, Phoridae, and Cecidomyiidae. The predominant group familiar to most mushroom growers is the pyemotid or red pepper mites. Mushroom growers have typically thought of nematodes as being harmful to the crop. Parasitic and saprophytic nematodes are associated with commercial mushroom yield loss. Beneficial nematodes, on the other hand, control insect populations.

Damage

- The mushrooms appear glossy and light brown, and small carpophore may be completely perforated and results in tissues crumble, when picked.
- In this, little direct damage from gnat will be evident on first-flush mushrooms as fly population has not developed sufficiently.
- The second and subsequent flushes may show some damage from larval feeding.

Management Strategies

- The movement of sciarid flies from one production room to another or from one mushroom farm to another is done mainly by unassisted adult dispersal and persistence of insects, which will crawl through any crack or crevice into the mushroom production facility.
- In temperate climates in winter with outside temperatures below 7°C, sciarid populations are lowest since migration and outside reproduction is limited due to temperature.
- Although the threat of fly problems is minimal at this temperature, sciarids can move from one production room to another by corridors, lofts, mezzanines, or outside the building; and they have been seen on the surface of snow 7–10 m away from production rooms where they have emerged.
- During warmer months, they will move from building to building, and to farms several kilometers distant.
- Under conditions of poor sanitation and hygiene, eggs, larvae or flies can enter the room on equipment that was not cleaned from previous spawning or casing procedures.
- By leaving spawn overnight in fly-infested corridor, infestation of flies can be introduced in production room by spawn or boxes.
- Growers who obtain mushrooms from other growers run the risk of introducing fungus gnats along with diseases, if other farms have a fly problem.

Monitoring

- ❖ When an invasion occurs, size of initial population and predicting its future size are important considerations.
- ❖ Monitoring gnat populations provides this information, but factors like climate, disease, immature and adult insect populations and growing practices are important.
- ❖ All production rooms to have monitor regardless of farm system as some monitor locations in the room captures more flies than another.
- ❖ In single-zone shelf farm or in phase II tray system, monitor should be placed in room prior to the compost cool-down period when air temperature reaches 43°C.
- ❖ Monitoring is more challenging for multi-zone farms with phase II and spawn-run tunnels.
- ❖ If tunnel is totally enclosed in a building, which is the most desirable, a monitor should be placed in the work area.

Cultural Practices

- ❖ Rapid cool-down at the end of phase II reduces time available for fly invasion while higher temperatures occur in compost during spawn-run and after casing, compared to harvest period.
- ❖ Shortening the spawn and case-run prolongs fly emergence in cycle of crop's phenology.
- ❖ Fly problems occur if the spawn-run is short.
- ❖ Fly control at the end of a crop is as important as control during spawn-run.
- ❖ Growers should heat-treat the old compost at 60 to 65°C for 8–12 hours to kill flies at all stages of development; these conditions will also kill most disease-causing fungi and bacteria.
- ❖ A crop may have to be terminated earlier than schedule dictates to ensure that population of emerging flies can be controlled prior to its spread to other locations.
- ❖ Prior to adding heat, an adulticide should be applied to eliminate adult fly escapees.

Biological/biorational Control

- ❖ There are many larvicides available which are most efficacious and efficacy is affected by the life stage of the larva at application
- ❖ Bacillus thuringiensis var. israelensis and diflubenzuron tend to be more effective against younger larvae, while entomopathogenic nem-atodes, especially Steinernema feltiae, and methoprene are more effective against the older larvae.
- ❖ Timing a larvicide application based on monitoring the fly invasion will provide maximum effectiveness.
- ❖ Larvicides must be applied when larvae are susceptible which is true for insect growth regulator, methoprene.
- ❖ The location of larval grazing area must be anticipated.

❖ Sciarid larvae tend to move up in casing from compost or remain in the casing, so a compost drench prior to casing is not efficacious for invasion of flies after casing takes place.

Chemical Control

❖ Chemical insecticides typically target adults in production rooms, work areas, or surfaces where the flies will rest, swarm, and roost.

❖ Growers should also treat walls, door jambs, and the plastic cover that is placed over the compost after spawning.

❖ Adulticides in the form of aerosols or dusts should be applied when the action threshold is reached.

❖ The use of chemical insecticides can be an important part of gnat control on a farm, but growers should integrate their application with other practices.

2. Mites

Information:

❖ Today, mites, like many other pests, rarely appear on farms where all technological parameters at preparation of compost and mushroom cultivation and also where strict hygiene is maintained.

❖ The mites which encounters during cultivation process, only few will counted as mushroom pests.

❖ Some mites feed on fungi and bacteria that exist in great amounts in decaying organic materials. The other members of this type of insects are predators that mainly feed on nematodes and smaller mites.

❖ In initial stage of composting, large populations of various mites might be found in the substrate, hence there's nutrition for practically all species on this stage.

❖ During phase 1 compost preparation process and its pasteurization, all mites die, hence the appearance of mites on farm is indication of either poorly prepared compost, or its inadequate pasteurization.

❖ Mites easily get in the growing rooms if straw, manure or spent compost and mushroom wastes are kept very close to farm.

❖ Under unfavorable conditions, some kinds of mites are able to take specific shape and remain this way for long period of time.

❖ In hypopus state, mite's body flattens and gets covered with thick shell that protects it from exposure.

❖ Using suckers, hypopus can adhere to moving objects like flies or rodents and immigrate that way.

❖ When conditions are favorable for vital activity of mites, hypopus slough taking their usual appearance.

❖ Mites can get inside the growing rooms not only with flies, but by adhering to personnel's clothes and footwear and/or implements.

- Mites of Tarsonemus myceliophagus species that are considered mushroom pests are so small, that they are not seen with a naked eye and such shiny, light brown in color mites feed on the mushroom hyphae and hyphae of many other competing fungi.
- Being present in great amount in the substrate, mites gnaw the bases of mushroom stipes, which in result become rounded and obtain a reddish-brown shade.
- Despite the fact that unlike the majority of mites and insects, the mites of Tarsonemus myceliophagus species reproduce rather slowly, the amount of damage they bring to the growing crop will depend on the time and degree of infection.
- If large population of mites enters the batch at spawning, the consequences of their activities can be rather significant.
- Red pepper mites got their names due to outward appearance of their reddish-brown accumulations, moving on mushroom caps or the casing surface.

Red pepper mite

Information:

- These mites feed on competing fungi, hence their presence shows poorly prepared compost.
- This kind of mites reproduce very fast, one female can lay up to 160 eggs within 5 days and even though pepper mites are not mushroom pests, they promote spreading of green molds not only in the infected room, but also in the whole farm.
- Being present in large amounts, they worsen the mushrooms' marketable appearance and bring unpleasant sensations and sometimes even cause allergy for mushroom pickers at work.
- In order to decrease the chances for pepper mites to spread trichoderma, many mushroom growers treat infected areas with gas burners and salt or tissue papers soaked in disinfectant.
- Colorless and translucent saprophytic mites of Tyrophagus spp., Caloglyphus spp. and Histiostoma spp. species can sometimes be seen on mushrooms.
- Unlike the species of mites described below, these mites are slightly bigger in size and slow-moving.
- Since mites feed on decaying vegetative remains, they only can be found in poorly prepared compost that isn't well suited for mushroom mycelium growth.

Prevention:

The following are the general measures which should be taken by mushroom growers to prevent the appearance of all kinds of pests during the mushroom cultivation process which includes:

- high-quality composting and compost pasteurization for obtaining selective medium for mushroom mycelium;
- maintaining strict hygiene on farm;
- controlling flies and other mite spreaders;

- ❖ cleaning of machinery, equipment and rooms that were used during spawning process and spawn run;
- ❖ steaming rooms in the end of cultivation cycle at necessary temperature;
- ❖ using disinfectants to clean all working surfaces and implements;
- ❖ immediate removal of spent compost from the farm

3. Nematode Pests

Information:

- ❖ Mushrooms nematodes can be grouped in two categories: beneficial and harmful.
- ❖ Mushroom growers have typically thought of nematodes as being harmful to the crop.
- ❖ Parasitic and saprophytic nematodes are linked with commercial mushroom yield loss.
- ❖ The occurrence of mycophagous nematodes in mushroom culture is rare under modern mushroom farming techniques, while saprophytic nematodes are common.
- ❖ The overall economic impact on production in nematodes is minimal; however, under heavy populations significant individual crop reduction is experienced.
- ❖ Beneficial nematodes control insect populations which may be either natural to population or introduced to mushroom crop to control mushroom pests.
- ❖ The nematodes encountered in mushroom compost/beds are categorized as myceliophagous, saprophagous predaceous, plant parasites and animal parasites on basis of their nature of parasitism.
- ❖ Myceliophagous and saprophagous forms hold significance as they effect the crop yield through direct feeding or indirectly.
- ❖ Predeaceous forms usually occur in low counts and are beneficial as they feed upon the harmful myceliophagous and saprophagous nematodes.
- ❖ Plant and animal parasitic nematodes rarely prevail in small numbers, merely as contaminants.
- ❖ Myceliophagous nematodes have often been observed to infest the mushroom beds and reduce productivity of compost and its market value.
- ❖ Nematodes have often been observed to infest the mushroom beds and reduce productivity of compost and its market value.
- ❖ Nematode infestation in mushroom is unique in a way that the mushroom nematodes have not only adapted themselves fully with ecological requirements of the crop and multiply very fast to occur in high number in mushroom beds, but these are only examples of ectoparasitic forms, capable of inflicting 100 per cent crop losses.
- ❖ Myceliophagous nematodes are known to be the most damaging as they have needle like stylet on their mouth part, with which they penetrate and puncture the mycelial cell and can suck the sap and leaves cell devitalized.

- After feeding upon one cell, the nematode shifts to another with the help of moist film present in the compost.
- The cell sap oozed out due to leakage from the damaged hyphae may also assist the nematode movement from one cell to another.
- The myceliophagous nematode affected beds show various types of symptoms like mycelium disappearance, browning and stunting of water soaked pinheads and the fruit bodies appear in patches.
- In this, compost emits a particular odour and yield declines drastically.

Control:

- Maintain hygienic conditions in growing rooms by using doors fitted with fly proof wire mesh and complete sanitation around the mushroom farm.
- Peat heat temperature during pasteurization should be maintained at 58-59°C for 4 hours.
- Casing soil should be sterilized properly by steam heat of around 65°C or with formatin.
- Wooden trays should be sterilized with steam, formalin or dipped in hot water at around 80°C for 2-3 minutes.
- Growing of nematode-resistant strains in rotation with crop of common white button mushroom helps in reducing inoculum level of nematodes in mushroom house.
- Nematode trapping fungi like Arthrohotrys spp. can be used as biological control agents for controlling mushroom nematodes where fungus traps the nearby nematodes and kills them.

6.4 Nematodes of mushroom

6.4.1 Information:

- The nematodes of mushroom are Ditylenchus Spp. (Tylenchida: Anguinidae) and Aphelenchoides Spp. (Aphelenehida: Aphelenchoididae):
- Nematodes encountered in mushroom compost/beds have categorized as myceliophagous, saprophagous predaceous, plant parasites and animal parasites, based on their nature of parasitism.
- Among nematodes, myceliophagous and saprophagous forms hold significance as they effect the crop yield by direct or indirect feeding.
- Predeaceous forms occur in low counts and are beneficial as they feed on harmful myceliophagous and saprophagous nematodes.
- Plant and animal parasitic nematodes rarely prevail in small numbers, merely as contaminants.
- Myceliophagous nematodes have often been observed to infest the mushroom beds and reduce productivity of compost and its market value.

- ❖ Nematodes observed to infest the mushroom beds and reduce productivity of compost and market value.
- ❖ Nematode infestation in mushroom not only adapt themselves fully with ecological requirements of crop and multiply very fast to occur in high number in mushroom beds, but are only examples of ectoparasitic forms, that inflict 100% crop losses.
- ❖ Nematodes are small microscopic thread like organisms which can be found in almost any type of environment, all types of soils and areas, where organic matter is found and are capable of rapid multiplication.
- ❖ The myceliophagous nematodes suck the mycelial sap and leave it devitalized.
- ❖ In mushroom cultivation , six genera, two belonging to order Tylenchida (Ditylenchulus spp. and Pseudhalenchus spp.) and four belonging to order Aphelenchida (Aphelenchoides spp., Aphelenchus avenae, Paraphelenchus spp. and Scinura spp.) have been found.
- ❖ Among tylenchids, predominant is Ditylenchus myceliophagous Goodey, and among aphelenchids, Aphelenchoides with 20 species reported from mushroom farms located globally are the most significant species, from these, 10 species have been reported from India.

6.4.2 Life-Cycle:

- ❖ Aphelenchids, shows mycelophagous behavior with short life cycle of 7-12 days having temperature as main role.
- ❖ composticola took 18 days to complete its life-cycle at 13°C; it was completed in 10 and 8 days at 18 and 23°C, respectively.
- ❖ The mean sex ratio was found to be 0.69 and was always less than one, the number of females being always more than males.
- ❖ The developmental period of A. sacchari from egg to egg has been reported to be 12 days where longevity of females and males was 35-80 and 35-100 days.
- ❖ agaricus completes one life-cycle in 8 days at 28°C which is a bisexual species with an average female to male sex ratio of 1.5: 1.
- ❖ In this, average life span of females and males, with sufficient food availability was 89 and 100 days, but females in water survived between 19-32 days.

6.4.3 Damage:

- ❖ Myceliophagous nematodes are known to be the most damaging as they have needle like stylet on their mouth part, with which they penetrate and puncture the mycelial cell.
- ❖ Such nematodes suck the sap and leave the cell devitalized and after feeding on one cell, the nematode shifts to another using moist film which is present in compost.
- ❖ In this, cell sap oozed out due to leakage from the damaged hyphae and assist nematode movement from one cell to another.

- The myceliophagous nematode affected beds show various types of symptoms like mycelium disappearance, browning and stunting of water soaked pinheads and the fruit bodies appear in patches.
- The compost emits a particular odour and yield declines drastically.

6.4.4 Control:

- Maintaining hygienic conditions in growing rooms by providing doors fitted with fly proof wire mesh and complete sanitation around the mushroom farm.
- The peat heat temperature during pasteurization must be maintained at 58-59°C for four hours.
- The casing soil should be sterilized properly either with steam heat (65°C) or with formatin.
- Wooden trays should be sterilized properly either with steam, formalin or dipped in hot water (80°C) for 2-3 minutes.
- Growing of nematode-resistant strains in rotation with the crop of common white button mushroom helps to reduce the inoculum level of the nematodes in the mushroom house.
- The nematode trapping fungi like Arthrohotrys spp. can be used as biological control agents for controlling mushroom nematodes. The fungus traps the nearby nematodes and kills them.

6.5 Symptomatology of nematodes and their managements

6.5.1 Information

- Mushrooms in India are being cultivated under semi-scientific conditions with limited inputs.
- As there are many nematode control methods that have carried out in Indian conditions, as maintenance of hygienic production, prophylactic and usage of plant products remained successful in minimizing nematode damage to mushrooms and giving maximum yields.
- Problems of nematode in mushroom are different where not only nematodes have adapted themselves fully as per ecological needs of crop, but also they tend to multiply very fast by inflicting 100% crop losses.
- In India, many mushroom cultivation and production works on small scale with small/marginal farmers or by woman, in unhygienic or subhygienic environment.
- With an increase in demand for fresh and processed edible and medicinal mushrooms in domestic and export market, requirement for scientific and commercially viable mushroom production technologies results in growing.

6.5.2 Grouping

Nematodes linked with mushrooms can be grouped as:

- myceliophagous nematode
- saprophagous nematode

6.5.3 Nematodes in Commercial Mushroom Production

Parasitic and saprophytic species of nematodes linked with commercial mushroom growing are routine inhabitants of most agricultural soils. Parasitic nematodes possess needlelike mouth parts that draws out the contents of mushroom cell. The mouth of saprophytic nematode is built for chewing. Saprophytic nematodes ingest bacteria, protozoa, and fungal spores, but do not attack the mushroom mycelium. Large populations of saprophytic nematodes may produce by-products which adversely affect mushroom spawn growth, but the extent of this is still to be proven. One fact regarding nematodes is certain, if present in compost, are survived Phase II pasteurization or over-cropped in woodwork of growing room. Pasteurization of compost and wood is main in mushroom-farm hygiene, so presence of nematodes suggests failure in sanitation procedures.

Saprophytic nematodes have been repeatedly collected during certain time period which were associated with disappearance of spawn and breakdown of compost in black soggy mass. Such nematodes form into moist glistening swarms that appears as upright cylindrical aggregates swaying to and fro on surface of casing. Such swarms can be seen without magnification by directing a beam of light at angle 45" to casing, where motion of swarms makes them obvious.

Nematodes are transported to crop by infested compost, peat moss and soil, that were transported on dust particles where mushroom flies occurs. Transportation from infested to noninfested areas occurs on tools, on clothing and hands of workers, and on flies. The nematodes will over-crop in inadequately pasteurized wooden boards and survive in compost inadequately pasteurized during Phase II. Saprophytic nematodes multiplies in ceilings of wood or concrete and in ceiling insulation. Once ceiling is infested, water droplets condenses on ceiling will have nematode larvae or their eggs, and crop becomes infested by falling droplets. A nematode infestation is sometimes accompanied by superficial sparse growth of fungus Arthrobotrys superba, a parasite of the nematode. Such mold appears as whitish cast on surface of casing in an area up to 4 or 5 feet in diameter.

6.5.4 Symptoms of nematode infestation

As mushroom growers are reluctant to disturb beds after casing, the early symptoms of nematodes attack are overlooked which results in less yield. Damage caused by nematodes shows following symptoms of nematode which appear in infected beds in succession:

- Mycelial growth is sparce, patchy and mycelium turns stingy
- Compost surface sinks
- Whiteness of spawn-run slowly changes to brown.
- Sporophore flushes are poor and delayed
- Alternate high and poor yield in successive flushes.
- Browning of pinheads.
- Decline in yield.
- Complete crop failure.

6.5.5 Measures for nematode management

- Cropping to be done in purposely built mushroom houses with proper ventilation.
- Strict hygienic and sanitation measures to be followed all through cropping period.
- Composting yard to be cemented which prevents direct contact of compost with soil.
- Composting yard to be disinfected with 4% formalin, 24 hours before compost preparation.
- All instruments, walls, floors and gallaries to be disinfected with 4% formalin.
- Composting ingredients should always be stored in clean area.
- Cleanliness to be maintained in and surroundings of mushroom farms.
- Casing mixture to be properly pasteurized.
- Manures used for composting to be thoroughly broken and its layers allowed to decay properly.
- In long method of composting, cover compost with double PVC sheet for 24 hours after third turning gives effective control of nematodes.
- Foot dips to be installed in front of each cropping room.
- Irrigation water should be clean.
- No person/worker to be allowed to enter in farm without proper disinfection of his/her hands and feet.
- All cropping rooms to be fly proof and only recommended insecticides should be sprayed for the control of flies.

6.6 Disorders of mushrooms

6.6.1 Information

- Plant problems are caused by living organisms like fungi, bacteria, viruses, nematodes, insects, mites, and animals.
- Disorders in mushroom can be biotic and abiotic.
- Abiotic disorders are caused by nonliving factors like drought stress, sunscald, freeze injury, wind injury, chemical injury, nutrient deficiency or improper cultural practices.
- Damage caused by various living and nonliving agents can appear very similar as with close observation of symptoms, accurate diagnosis can be difficult.
- When cause of plant health problem is not readily diagnosed, it's important to take a systematic approach and carefully consider site conditions, weather conditions, care of the plant, and the known biotic disease agents of that plant.
- The first important step is to determine the identity of the plant and its requirements for healthy growth.

6.6.2 Abiotic Disorders

- In mushrooms there are large number of abiotic agents which create unfavourable environment for proper growth, resulting into quantitative and qualitative loss.
- Abiotic agents can be low or high moisture in substrate, pH and temperature, CO_2 concentration in room, wind velocity, fumes and relative humidity.
- Such agents make the substrate nonselective for mushroom mycelium and encourage other moulds and pests while some interfere with normal mushroom production.
- Management of environment has great importance in mushroom cultivation and any deviation from optimum requirements may lead to various kinds of abnormalities.

The following abiotic disorders are observed in mushrooms:

Stroma

- Stroma are noticeable aggregations of mushroom mycelium on surface of spawned compost or the casing. Discrete aerial patches of white mycelium form a dense tissue layer on the substrate surface.
- Stroma can be easily peeled from the surface of compost or casing. Stroma form on the compost in small localized areas and the smaller patches can coalesce into larger areas.
- After casing, stroma may form on the casing above a patch of compost-borne stroma or on casing where stroma does not exist in the compost. Stroma on casing develops in advance of pinning but rapidly putrefies once watering begins. Mushrooms can develop on stroma, but this is somewhat unusual.
- A sector is a portion of spawn that is distinctive when compared to the general appearance of spawn. A sector may be extra-white, extra dense or extra-ordinary fluffy and is always different from the normal spawn. Sectors appear on or in the compost and on the casing, and tend to disappear as the crop ages.
- Removing patches of stroma does not ensure growth of mushrooms in these areas, so removal of stroma is a matter for each farmer to decide. This disorder has been commonly observed in seasonal farms in HP where proper aeration is lacking.

Weepers

- Mushrooms described as being 'Weepers' typically exude considerable amount of water from mushroom cap.
- When small water droplets exude from stem or cap, the mushrooms are called leakers. These water droplets may be few in number and relatively isolated from each other or may be sufficiently numerous to cover the mushrooms.
- The distinction between a 'leaker' and 'weeper' is that the water droplets remain as droplets on the leaker mushrooms while it actually falls or flows from a weeper.
- Weepers are usually noticed since they are quite unusual. A weeping mushroom can dissolve into a white foam. Water collects on the casing surface beneath a weeper and the area develops a putrid odour becoming a 'stinker'.

- Factors that induce a mushroom to become a weeper are not known but low-moisture compost-less than 64% coupled with high moisture casing is where weepers are frequently seen.
- The combination of these two conditions often foster weeper mushrooms prior to and during the first break. Smooth white mushrooms seems to have some sort of protection against leakers and weepers. Other major types-off-white, cream, golden white are susceptible to this malady.

Flock, Hard Cap And Open Veil

- Flock is a physiologically induced malformation of the mushroom's cap and gill tissue.
- The cap opens pre-maturely and the gills of the affected mushrooms are rudimentary, poorly developed and have little pigmentation.
- The flocked mushrooms generally appear in first flush and may disappear in subsequent flushes but in some cases it continues increasing in subsequent flushes.
- Environmental conditions including diesel exhaust, oil-based point fumes and certain anticorrosive chemicals in steam boilers or certain diseases like dieback, brown plaster mould and false truffle induce flock symptoms.

Hollow core and Brown pith

- These two disorders seem to afflict cream strains much more than other strains, although off-white strains can have hollow core.
- When the bottoms of the stems are trimmed after harvesting, a circular gap is seen in the centre of the stem.
- This hole may extend the length of the stipe or it may be shorter.
- When the hollow cut end portion is brown in colour the sale price is considerably reduced.
- This abnormality seems to be related to watering and water stress.

Purple stem

- Cut stems of the mushrooms develop a deep purple colour within few hours of harvest or after being in cold storage (36°F) overnight.
- At times colour is closer to black than purple and it occurs in all strains smooth white, off-white, cream and brown.
- Generally mushrooms from 3rd break to the end of the crop are most susceptible.
- Polyphenol oxidase, an enzyme increases in later-break mushrooms and this enzyme influences pigment formation.
- Conditions that predispose mushrooms to this phenomenon are unknown but the frequency and the amount of water applied before harvest seems to affect its occurrence.

Rose Comb

- Large lumps and swelling are visible on the mushroom cap.
- The gills often grow in the top of the cap tissue and even on the top of the cap.
- These misshapen gills make the swellings look spongy.
- The mushrooms can even burst or split and then turn brown.
- The abnormality is caused by gases or vapours coming from solvents, paint or oil products and polluted casing soil.

Scales or crocodiles

- Scales arise through the surface tissue failing to grow while the cap develops further.
- The main reason for scales being formed is poor climate control, in particular too much drying out or too great air velocities.
- Strong vapours of formaldehyde or pest-control products in excess can also cause the outer layer of the skin of half-grow mushrooms to tear off.
- As the mushroom continues to grow, the skin bursts and so-called 'crocodile' skin is formed.
- The off-white and cream mushroom strains are more sensitive to scales than white mushrooms.
- This is the most common and serious malady affecting button mushroom in seasonal farms in HP.

Long stemmed mushrooms

- The presence of long stems in combination with a number of other symptoms can indicate virus diseases but it is often the result of too high CO2 concentration so that the stems extend more (drumsticks).
- With the improvement of aeration such conditions can be avoided.

Brown Discolouration

- Browning of small pin heads or half grown mushrooms is very common on seasonal mushroom farms.
- This may be caused by high temperature, sprinkling at high water pressure (maximum pressure is 0.4 atm), chlorinating with too high a chlorine rate [maximum rate is 500ml (10%) per 100 litre of water per 100m^2] or incorrect use of formalin, e.g. by spraying the mushrooms with a formalin solution.

Unit 7

Post Harvest Processing of Mushroom

7.1 Post harvest care of mushroom

7.1.1 Information

- Mushrooms are rich source of good quality protein, having most of essential amino acids, minerals and vitamins with low calories.
- Production of mushroom is done all through the year by environmentally controlled units, but seasonal growers come in play during winters and supply at local market causing less profit due to fall in price and spoilage which comes due to market surplus.
- Mushrooms are highly perishable and get spoiled due to browning, wilting, liquefaction, loss of texture, aroma, flavour, etc, making it unsaleable.
- Most of mushrooms, being high in moisture and delicate in texture, cannot be stored for more than 24 hours at ambient conditions prevailing in the tropics.

7.1.2 Post Harvesting of Mushroom

The objective of post-harvesting is to determine a longer storage period and to upkeep a high quality for mushroom products. Once the fruiting body of mushroom matures, deterioration begins with formation of brown colouration and so quality deteriorates which lead to market loss. To overcome such problem, specially during peak season, suitable post harvest management/practices are to be followed to increase shelf life and marketability which involves following processes:

- Harvesting
- Grading
- Pre Cooling: Fresh or Processed
- Packing: POLYPOUCHES CARTONS
- Transportation: ROAD or AIR
- Marketing: DOMESTIC TRADE or EXPORT

Harvesting:

- Harvesting results in severing the supply of substrates by mycelium for growth and respiration despite of fact that mushrooms after harvest continues to grow with development of gills and formation of basidiospores.
- In this, metabolic shift takes place which result in compositional changes.
- The shorter shelf life of mushrooms is due to its very high respiration rate-of about 28.2-43.6 mg CO_2 per kg fresh weight per hour at 0°C and 280 mg CO_2 per kg fresh weight per hour at 19°C.
- The respiratory heat in mushrooms is reported to be 14 kcal/kg/24 h at 20°C and mannitol acts as preferred post harvest respiratory substrate.
- Mushrooms are generally harvested after 3 weeks of casing.
- Button mushrooms harvest when the CAP size is 30-45 mm in diameter.
- Oyster mushroom is harvested when fruiting body becomes curled under edges and well formed gills.

Grading

- Mushroom grading is done to market the harvested mushrooms which is done by segregating mushrooms in different grade standards as per market demand.
- The mushrooms are graded as A, B and C as per Directorate of Marketing & Inspection.

Pre-cooling

- The temperature of mushroom after picking varies from 15-18°C which rises steadily during storage, as a result of respiration and atmospheric temperature.

- ❖ With this, heat results in deterioration of mushroom quality, so heat should be removed once mushrooom is harvested and its temperature to be brought down to 4-5°C.
- ❖ Precooling is important in post-harvest stage of perishable-produce industry where heat is removed from freshly harvested fruits and vegetables to reduce the loss in quality of produce once it's been picked.
- ❖ Precooling increases the shelf-life of fresh produce, which results in more profits to produce growers.
- ❖ Cooling system will depends on quantity of mushroom, which can be refrigerator or cold room.
- ❖ In pre-cooling, harvested mushroom is kept in plastic bag and are stored in cooling unit.
- ❖ Vacuum cooling is cooling system where water existing in cell walls and inter hyphal spaces of produce is evaporated under pressure which lowers the temperature.
- ❖ Evaporative cooling, hydro-cooling, forced-chill air, ice bank or vacuum cooling systems, mushrooms can be cooled upto 2-4°C.
- ❖ In this, temperature of mushrooms increases by respiration after picking and respiratory rate increases with increase in storage temperature.
- ❖ Mushrooms at temperature 10°C have 3.5 times higher respiratory capacity than those at temperature 0°C, which shows immediate shifting of mushrooms to refrigerated zone.

Air-cooling	• Air velocity at least 60 m/min • Mushrooms usually placed in boxes, various packages or punnets and cooled within 24 h, in a rate of 0.5°C/h, but unevenly • Air is usually dehumidified during refrigeration, so the vapor pressure of the cooled air is low and dehydrates the mushroom surface • There is about 3% weight loss in cooling from ambient temperature to 2°C in unwrapped mushrooms that reach quicker the store temperature particularly if they are placed on the top of the pile • The use of films (e.g., PVC) in mushroom packaging is essential for preventing great weight losses even if it hampers the quick cooling
Forced cooling	• Higher air pressure is developed on one side of the packages than the other and air is directed through the produce cooling it rapidly • It is accomplished by using three systems cold-wall, forced air tunnel, and serpentine cooling • A spray-moist chiller can be also used for cooling mushrooms rapidly in an hour at 16°C–18°C, without moisture loss
Ice bank cooling	• Air used to cool mushrooms is almost saturated as it has previously passed through ice-cold water, cooled by direct heat exchange with it and filled with vapor, a system preventing mushrooms from water loss
Vacuum cooling	• It is a rapid and uniform (temperature reduction from ambient to 2°C in 15–25 min), yet expensive method • Subjecting mushrooms at very low pressures water from their surface is evaporated resulting in lower temperatures • The color of vacuum-cooled mushrooms is superior to conventional cooled mushrooms, but the weight losses reported are 1% per 5°C of drop, much greater than those of forced dry or wet air • No differences in quality or hyphal structure were detected between vacuum and conventionally cooled mushrooms • Overwrapping with perforated films is essential for maintaining the appearance of vacuum cooled mushrooms

Packaging

- ❖ Packaging plays an important role in the control, marketing and consumption of the yield and production of mushrooms.
- ❖ It protects the quality during storage and transportation for retail sale and storage for consumers.
- ❖ The packaging of mushroom from farm till it reaches the consumers is an important aspect in post-harvest control.
- ❖ Before packaging of mushroom, grower should clean the mushroom and then put it into thin polypropylene bags that should be 5% perforated for ventilation.
- ❖ Mushrooms are generally packed in polythene bags of 250 gm-400 gms for local markets, while for long distances, they are packed in pulp board punnets wrapped with PVC films.
- ❖ Precooled mushrooms are packed in insulated containers having ice, so that they remain fresh, healthy during long transport.
- ❖ Mushrooms used for longer distant, are packed in CFB cartons by refrigerated trucks while mushrooms for shorter distance markets uses wooden boxes with crushed ice in polypacks.
- ❖ Mushrooms are packed in polypouches for local retailer's that are dispatched by cycle, cycle rickshaws, auto-rickshaws etc.

Transport

- ❖ The effect of pre-cooling and packing will is partially negated when product is later stored and transported in hot environment.
- ❖ Mushrooms need refrigerated transport, to keep the mushrooms cool during transport.

- For short distance transportation, the polypacks of mushrooms are stacked in small wooden cases or boxes with sufficient crushed ice in polypacks.
- For long distance transport of large quantities of mushrooms, refrigerated trucks are used.

Preservation of Mushroom

- Mushrooms are being preserved in order to extend their shelf life.
- Fresh mushrooms only keep for about five days after picking while preserved mushrooms can be kept for longer period of time.
- Once mushrooms are picked from the farm, they are transported to processing factories and are immediately kept in vacuum where they suck up lot of water.
- Freshly harvested mushroom is highly perishable as it is susceptible to deterioration by enzyme and microorganisms.
- Merely producing mushroom is of no use unless these are properly preserved, keeping in view the export objectives, so proper processing methods should be used which includes short term and long term preservations.
- Canning, pickling, and drying processes are used for long-term storage as such processes are not always suitable for all types of mushrooms.
- In such process, quality of finished product is rarely comparable with that of fresh mushrooms.
- Canning is a common process for preserving mushrooms where cleaned mushrooms are kept in cans with 2.5 % sodium chloride and 0.25–0.5% citric acid.
- In canning, cans are sealed and sterilized in autoclave for one hour at 100-120°C.
- Mushrooms can be dried by sun drying and thermal power drying where picked mushrooms are exposed to sun for 2-4 days.
- Dried mushrooms are highly hygroscopic and readily absorb moisture from the air, so they should be kept in polyethylene bags, sealed and kept in dry, cool and dark place.
- For prolonged storage, dried mushrooms should be packed in cartons or wooden boxes and kept at 2-5°C in a low temperature store.

7.2 Processing methods of mushrooms

7.2.1 Information

Mushroom is nutritious with many varieties having many vitamins and minerals with low sugar and fat content. It can be grown in artificially created and controlled environment where temperature ranges between 20C-30 C with relative humidity of around 55%-75%. Mushrooms can be ideally grown in hilly areas or in artificially controlled sheds in plains. Fresh harvested mushrooms can be readily sold in market after processing and drying.

The mushroom has been processed in various products to increase its consumption, providing the health and nutritional benefit to mankind. Fresh mushrooms have limited life, so processing is required to improve the shelf life. For this, initially harvested mushrooms are washed in cold water and then blanched in boiling water for 3-4 minutes which can be dehydrated in drier and then finally packed. It is advisable to pre treat mushroom in brine solution to prevent discoloration. Packing is very crucial as formation of moisture contaminates mushroom very quickly. The yield depends on many factors as moisture content in fresh mushroom, type of drier , process employed, moisture content require in the finished product etc.

Mushrooms are known for their ability to grow rapidly, often seemingly overnight. Unfortunately the same can almost be said about their deterioration, with most mushroom varieties beginning to decay rapidly after harvest unless they are properly cooled and stored. Unfortunately even with careful care they still have a relatively short shelf life compared to many other types of produce.

Processing of mushroom involves following processing methods:

Processing for Short-Term Preservation

1 Cooling

There are certain methods that are used for cooling of mushrooms, so that heat can be removed which brings mushrooms' temperature down to proper levels which includes:

Hydro-cooling:

- It involves rapidly submerging the mushrooms in near-freezing water.
- It is extremely effective at removing initial field and harvest heat so that the mushrooms can be cooled to final storage temperatures.

Forced-Air Cooling:

- It involves placing the mushrooms in refrigerated room like cooler and arranging them around a fan or fans which then pull cool air through them.
- Such method results in much faster cooling than room cooling alone, which is often not sufficient to cool the mushrooms quickly and thoroughly enough.

Vacuum Cooling:

- It is the process of exposing moist mushrooms to extremely low pressure levels.
- Such methods results in water vaporizing and absorbs latent heat, rapidly and uniformly cool the mushrooms.

2 Minimal Processing

Mushrooms are consumed for medical purpose also through food. Due to its short shell life and perishability, it cannot bc stored for longer period. With this, grower/market seller should give more attention in terms of minimal processing, as with this, mushroom shelf-life can be increased using drying and canning operations.

3 Effect of Packaging on Mushroom Shelf-Life

- Clean fresh mushrooms are packed in different ways as per wholesale, retail and transport needs.

- Such mushrooms can be sold as loose or in trays/punnets overwrapped with plastic.
- Macro perforated PP films that limit dehydration without modifying atmosphere in package are used, but over perforation leads to excessive mushroom water loss resulting into wrinkling and occurrence of brown patches on its surfaces.
- Proper packaging is important for mushroom, as it protects from bruises throughout transportation and extends shelf life by reducing desiccation and vapor condensation.
- Design, color and labeling on packed mushrooms shows impression about mushroom quality, whereas plastic films available nowadays shows low gas permeability.

Processing for Long-Term Preservation

1 Freezing

2 Canning

3 Drying

Freezing

- It is a famous method of mushroom preservation which gives storage stability and allows mushroom consumption all through the year.
- Such method gives consumers a product having high nutritional value and quality attributes.
- It is seen that all mushroom species can be consumed as frozen products.
- Mushroom freezing process includes mushroom cleaning, washing, cutting, grading/ sorting and blanching.
- Freezing mushrooms is a great storage technique as fresh mushrooms last for only week in the refrigerator, so freezing allows grower to save for longer period of time.
- Blast freezing is common method used in mushrooms as blast freezers rapidly bring the temperature of foods down, freezing them extremely quickly from −25°C to −30°C, creating small ice crystals that damage less mushroom cells.
- Once mushrooms has frozen, they can be moved to conventional freezer for storage, as long as the freezer stays cold enough to keep them frozen.

Canning

- Canning is a technique by which mushrooms can be stored for longer periods.
- Caning process comprises of cleaning, blanching, filling, sterilization, cooling, labelling and packaging operations.
- For good quality canned mushrooms, processing should be done as soon as possible after the harvest and in case of delay; mushrooms can be kept at 4-5°C till it get processed.
- Mushrooms having stem length of 1 cm are preferred and are canned whole, sliced and stems and pieces as required.
- Well graded fresh mushrooms white in color, without dark marks on either caps or stems are preferred for canning.

- The whole mushrooms is washed 3-4 times in cold running water to remove adhering substances on it.
- Blanching of mushroom is done to inhibit polyphenol oxidize enzymes activity and to inactivate microorganisms and in removing gases from mushroom tissue and reduction in bacterial counts.
- The blanching can be done in stainless steel kettles filled with boiling solution of 0.1% citric acid and 1% common salt.
- The blanching time ranges from 5-6 minutes at 95-100°C.
- Mushrooms after blanching are filled in sterilized tin cans.
- Brine solution is added to mushroom-filled cans after bringing its temperature to 90°C.
- Once the can is filled, it is exhausted by passing in exhaust box for 10-15 minutes, so that temperature in centre of can reaches to 85°C.
- Further, cans are sealed hermetically with double seamier and is kept in upside down position.
- Once the can is exhausted, sterilization of can is required where can is heated up to 118°C to prevent spoilage by microorganisms during storage.
- The can gets cooled down immediately after sterilization process in order to stop over-cooking and to prevent stack burning.

Drying

- It is simple and reliable method of preserving mushrooms.
- In this, mushrooms is sliced for fast processing.
- Drying is based on principle that water activity of product is reduced at specific level so that it is microbiologically and physicochemically stable

7.3 Spent mushroom

7.3.1 Information

- Spent mushroom substrate is soil-like material which remains after a crop of mushrooms.
- Spent substrate is high in organic matter making it desirable for use as soil amendment or soil conditioner.

- Substrate prepared specifically for growing mushrooms is blend of natural products which covers ingredients such as wheat straw bedding with horse manure, hay, corn cobs, cotton seed hulls, poultry manure, brewer's grain, cotton seed meal, cocoa bean hulls and gypsum.
- Growers may add ground soybeans or seed meal supplements later in production cycle.
- On top of substrate, farmers apply casing layer, which is a mixture of peat moss and ground limestone where casing material provides support for the growing mushrooms.
- Spent mushroom substrate has some nutrients available for mushroom, but it is more economical to replace the substrate and start with new crop.
- Before removing spent substrate from mushroom house, grower pasteurizes it with steam to kill any pests or pathogens that may be present in substrate and casing.
- Final pasteurization will kills weed seeds, insects, and organisms that causes mushroom diseases. Grower/users may consider spent substrate clean of weed seeds and insects.
- Mushroom growers apply registered pesticide during the crop cycle while local garden center sells most of same pesticides which mushroom farmer uses.
- Even if pesticides have been applied, they are generally hard to find as organic matter in substrate binds pesticides and these compounds decompose rapidly at high temperatures which are used for pasteurizing complete crop.
- The pesticide residue on spent substrate is low, so some farms are organic and will not use chemical pesticides.

7.3.2 Characteristics

The typical composition of spent mushroom substrate directly from mushroom house varies slightly. As raw materials and other cultural practices change, each load of fresh spent substrate has slightly different element and mineral analysis. Sometimes, fresh substrate is placed in fields for at least one winter season and then marketed as weathered mushroom soil. Such aged material has different characteristics and microbial activities in field result change in the composition and texture. In this, salt content changes during the aging period and if grower have any specific questions concerning characteristics of fresh or aged spent substrate, he/she needs to consult the local authorities.

7.3.3 Uses

- There are many appropriate uses for spent mushroom substrate such as:
- The spent mushroom substrate is good to spread on top of new seeded lawns as it provides cover against birds eating the seeds that holds the water in soil during seed germination.
- As some plants and garden vegetables are sensitive to high salt content in soils, the fresh spent substrate should not be used around such plants.

- Grower may use spent substrate weathered for 6 months or longer in all gardens and with most plants.
- Getting spent substrate in fall and winter, allow it to weather, will make it ready to use in a garden in following spring.
- Spring and summer are best time to use weathered material as mulch.
- As soil amends, spent substrate adds organic matter and structure to the soil.
- Spent substrate mainly improves soil structure and provide few nutrients.
- Spent substrate is the choice ingredient by certain companies who are into making potting mixtures, which are sold in supermarkets/garden centers.

Unit 8

Cultivation of Different Mushroom

8.1 Cultivation of oyster mushroom

8.1.1 Introduction

- Oyster mushroom belongs to class Basidiomycetes and family Agaricaceae, which in India is popularly known as dhingri.
- It is a basidiomycete and belongs to Genus `Pleurotus'.
- It grows naturally in temperate and tropical forests on dead and decaying wooden logs or on dying trunks of deciduous or coniferous woods or on decaying organic matter.
- The fruit bodies of the mushroom are distinctly shell or spatula shaped with different shades of white, cream, grey, yellow, pink or light brown depending on the species.
- Out of 38 oyster mushroom species, 25 species are commercially cultivated in different parts of world which includes P. ostreatus, P. flabellatus, P. florida, P. sajor-caju, P. sapidus, P. cystidiosus, P. eryngii,P. fossulatus, P. opuntiae, P. cornucopiae, P. yuccae, P. platypus, P. djamor, P. tuber-regium, P. australis, P. purpureo-olivaceus, P. populinus, P. levis, P. columbinus, P. membranaceus etc

8.1.2 Origin

- Cultivation of sp. of oyster mushroom was initiated on experimental basis in Germany during the year 1917 on tree stumps and wood logs.
- Growing technology was perfected in USA by Block, Tsao and Hau.
- Cultivation of different varieties of oyster mushroom in India was initiated in early sixties.
- Commercial cultivation of oyster mushroom in India began in mid-seventies.

8.1.3 Features

- Oyster mushrooms have three distinct parts; fleshy shell, short/long lateral or central stalk and long ridges and furrows below pileus.
- It is a suitable fungal organism which produces protein rich food from various agrowastes without composting.
- In this, gills stretches from the edge of cap down to stalk and bear the spores.
- In oyster mushroom, spores are smooth, cylindrical and germinate very easily on any mycological media in 48-96 hrs.
- The mycelium of Pleurotus is pure white in colour.

8.1.4 Production

- Oyster mushrooms are the third largest cultivated mushroom across the world.
- China is the world leader in Oyster production which contributes to around 85% of total world production.
- The present production of oyster mushroom crop in India is only around 1500 tonnes due to low domestic demand.
- Oyster mushrooms are grown from mycelium and propagate on base of steam-sterilized cereal grain

8.1.5 Cultivation

The procedure for oyster mushroom cultivation can be described in following steps:

- Preparation or procurement of spawn
- Substrate preparation
- Spawning of substrate
- Crop management

1 Preparation or procurement of spawn

Spawn

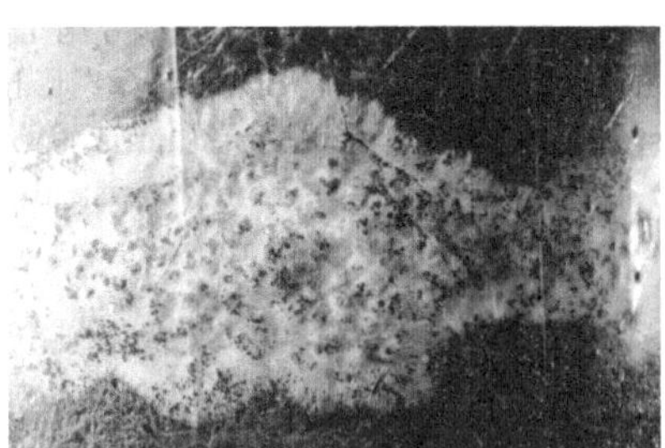

- Oyster mushrooms are grown from mycelium propagation on base of steam-sterilized cereal grain which is grain/mycelium mixture known as spawn.
- Spawn is used to seed the mushroom substrate.

- Spawning is a process where mushroom mycelium is introduced in compost which is carried out with some form of carrier that can be mixed in compost, fungus-coated cereal grains.
- Many spawn is made with mycelium from stored culture instead of mycelium whose parent was a spore as spores yield a new strain with unpredictable performance.
- Making of spawn is a complex task and is not feasible for common mushroom grower.
- Spawn of many oyster mushroom species can be purchased from commercial spawn makers who can provide instructions for its usage.
- Inoculum for spawn production is produced in polyethylene bags with microporous breather strip for gas exchange.
- Many commercial spawn production firms produces spawn only from inoculum that has met strict quality control standards.

Preparation

- A pure culture of Pleurotus sp. is needed for inoculation on sterilized substrate.
- It takes 10-15 days for mycelial growth on cereal grains.
- It has been reported that jowar and bajra grains are superior over wheat grains, as mycelium of oyster mushroom grows very fast on wheat grains and takes 25-30 days old spawn starts forming fruit bodies in bottle itself.
- Take 10kg of wheat grains and boil it for 15 minutes in 15 Liters of water and then allowed the wheat grains to soak for 15 minutes without heating.
- The excess water gets drained off and grains are cooled in sieves.
- The grains should be turned several times with spoon for quick cooling.
- The cooled grains are mixed with gypsum ($CaSO_4 \cdot 2H_2O$) and 30g of calcium carbonate ($CaCO_3$).
- The gypsum prevents the grains from sticking together and calcium carbonate is required to correct the pH of water.
- The prepared grains are filled in 1/2 liter bottles or polypropylene bags and autoclave for 2 hours at 121°C.
- Grower should see that after sterilization, the material should have a pH value of 7.
- The bottles are inoculated with grains or bits of agar medium colonized with mycelium, and incubated at 22-24°C in dark place.
- The mycelium completely spreads through the grains in about 2 weeks.

2 Substrate Preparations

- Oyster mushroom can be cultivated on a large number of agro-wastes having cellulose and lignin which helps in more enzyme production of cellulose that is correlated with more yield such as straw of paddy, wheat and ragi, stalk and leaves of maize, millets and cotton, used citronella leaf, sugarcane bag asse, saw dust, jute and cotton waste, dehulled corncobs, pea nut shells, dried grasses, sunflower stalks, used tea leaf waste, discarded waste paper and synthetic compost of button mushrooms etc.

- ❖ It can also be cultivated by using industrial wastes like paper mill sludges, coffee byproducts, tobacco waste, apple pomace etc.

The popular methods of substrate preparation are:

- ❖ Steam Pasteurization
- ❖ Hot Water Treatment;
- ❖ Sterile Technique (Till method);
- ❖ Fermentation or Composting;
- ❖ Chemical Sterilization

i. Steam pasteurization

- ❖ In this, pre-wetted straw is packed in wooden trays or boxes and are kept in pasteurization room at 58-620C for 4 hours.
- ❖ In this, temperature of pasteurization room is manipulated using steam through a boiler.
- ❖ Here, the substrate after cooling at room temperature gets seeded with spawn.
- ❖ The complete steam pasteurization process takes atleast 3-5 days.

ii. Hot water treatment

- ❖ In this, substrate which can be wheat straw after chopping gets soaked in hot water at around 65-70°C for 1 hrs and at 85°C for 30-45 minutes when using paddy straw .
- ❖ Once excess water drains out and cooled, spawn is added.
- ❖ Hot water treatment makes the hard substrate such as maize cobs, stems, etc. soft so that growth of mycelia takes place easily.
- ❖ Such method is not good for large-scale commercial cultivation.

iii. Chemical sterilization technique

- ❖ In this, different species of Trichoderma, Gliocladium, Penicillium, Aspergillus and Doratomycs spp. are competitors fungi on straw during oyster mushroom cultivation.
- ❖ If these are there on straw during spawn run, they do not allow growth of mushroom mycelium resulting in yield loss or complete crop failure.
- ❖ In this, if wheat straw or paddy straw is treated by steeping in chemical solution of carbendenzim 50% WP and formaldehyde for 16-18 h, then many competitor moulds are killed or growth gets suppressed for 25-40 days after spawning.

iv. Sterile technique

- ❖ It is called as Till method where chopped substrate after soaking in cold water is kept in heat resistant polypropylene bags and are sterilized in autoclave at 20 p.s.i. for 1-2 hours with spawning under aseptic conditions.
- ❖ Sterile technique is good for research work instead of large-scale commercial production.

v. Fermentation or composting

- It is the advancement of composting technique used for white button mushroom.
- It is good for hard substrates like cotton stalks, maize stalks, leguminous stubbles, etc.
- In this, aerobic and anaerobic fermentation of substrate is suitable for Pleurotus cultivation.
- Here composting to be carried out on covered area or shed.
- For doing this, the substrate gets chopped in 5-6 cm of long pieces and ammonium sulphate or urea and lime of around 0.5%-1% and 1% are added on dry weight basis.
- Grower can also add horse manure or chicken manure of around 10% dry weight instead of nitrogenous fertilizers.
- When lime is added, physical structure of the compost gets improved and after mixing the ingredients, sprinkle water till it gets completely wet.
- Now prepare triangular heap of 75-90 cm with not more than 1 meter height and after 2 days of fermentation, turn the pile by adding 1% superphosphate and 0.5% lime.
- With this, the compost gets ready after 2 days of turning.

3 Spawning of Substrate

- Freshly prepared grain spawn is best for spawning.
- Old spawn of around 3-6 months stored at room temperature 20-30^0 C forms very thick mat like structure due to mycelium aggregation and sometimes young pinheads and fruit bodies start developing in the spawn bottle itself.
- The spawning should be done in a pre-fumigated room (48hrs.with 2% formaldehyde).

4 Crop Management

(A) Incubation

- In this, spawned bags, trays or boxes are arranged in dark cropping room on raised platforms/shelves for mycelium colonization of substrate.
- Mycelium can grow from 10^0 to 33^0 C, but optimum temperature for spawn running lies between 22^0 to 26^0 C.

(B) Fruiting

- If mycelium fully colonized the substrate, then fungus is ready for fruiting.
- The contaminated bags with moulds can be discarded while bags with patchy mycelial growth may be left for few more days to complete mycelial growth.
- In this, many species require different temperature regimes with high humidity of around 70%-85% during fruiting.
- Frequent spraying of water is needed in cropping room as per the atmospheric humidity.
- Fruit body produced under humid conditions is bigger with less dry matter while those developed at 65-70% relative humidity are small with high dry matter.

- CO_2 concentration during cropping to be less than 600 ppm or 0.6%.
- Sufficient ventilation has to be provided during fruiting.

8.1.6 Importance

- The economic importance of the mushroom lies mainly in its use as food for human consumption.
- Mushrooms are rich in Vitamin C and B complex and protein content varies between 1.6 to 2.5% with most of mineral salts required by human body.
- The folic acid present in oyster mushrooms helps to cure anemia.
- It is suitable for people with hyper-tension, obesity and diabetes due to its low sodium: potassium ratio, starch, fat and calorific value.
- Alkaline ash and high fibre content makes them suitable for consumption for those having hyperacidity and constipation.
- A polycyclic aromatic compound pleurotin has been isolated from P. griseus which possess antibiotic properties.
- The spent straw can be re-cycled for growing oyster mushroom after supplementing with wheat or rice bran @ 10-15 % and also for preparing compost of white button mushroom after suitable supplementation with nitrogen rich horse or chicken manure.
- The spent straw can be used as cattle feed and for bio-gas production,
- In this, slurry can be used as manure.

8.2 Cultivation methods of oyster mushroom

8.2.1 Procedure

The procedure for oyster mushroom cultivation can be described in following steps:

(i) Preparation or procurement of spawn

(ii) Substrate preparation

(iii) Spawning of substrate

(iv) Crop management

1 Preparation or procurement of spawn

Spawn

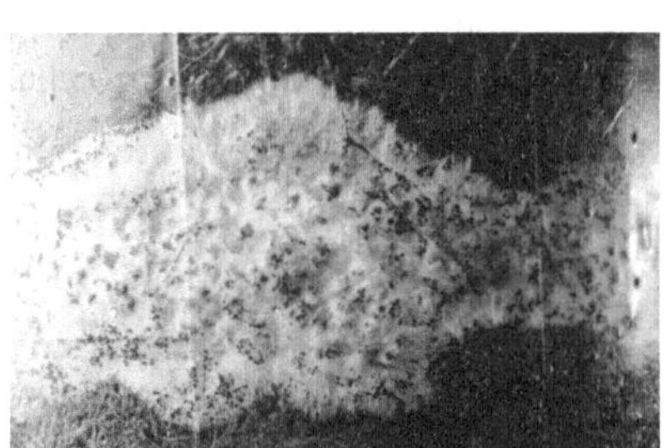

- ❖ Oyster mushrooms are grown from mycelium propagation on base of steam-sterilized cereal grain which is grain/mycelium mixture known as spawn.
- ❖ Spawn is used to seed the mushroom substrate.
- ❖ Spawning is a process where mushroom mycelium is introduced in compost which is carried out with some form of carrier that can be mixed in compost, fungus-coated cereal grains.
- ❖ Many spawn is made with mycelium from stored culture instead of mycelium whose parent was a spore as spores yield a new strain with unpredictable performance.
- ❖ Making of spawn is a complex task and is not feasible for common mushroom grower.
- ❖ Spawn of many oyster mushroom species can be purchased from commercial spawn makers who can provide instructions for its usage.
- ❖ Inoculum for spawn production is produced in polyethylene bags with microporous breather strip for gas exchange.
- ❖ Many commercial spawn production firms produces spawn only from inoculum that has met strict quality control standards.

Preparation

- ❖ A pure culture of Pleurotus sp. is needed for inoculation on sterilized substrate.
- ❖ It takes 10-15 days for mycelial growth on cereal grains.
- ❖ It has been reported that jowar and bajra grains are superior over wheat grains, as mycelium of oyster mushroom grows very fast on wheat grains and takes 25-30 days old spawn starts forming fruit bodies in bottle itself.
- ❖ Take 10kg of wheat grains and boil it for 15 minutes in 15 Liters of water and then allowed the wheat grains to soak for 15 minutes without heating.
- ❖ The excess water gets drained off and grains are cooled in sieves.
- ❖ The grains should be turned several times with spoon for quick cooling.
- ❖ The cooled grains are mixed with gypsum ($CaSO_4.2H_2O$) and 30g of calcium carbonate ($CaCO_3$).
- ❖ The gypsum prevents the grains from sticking together and calcium carbonate is required to correct the pH of water.
- ❖ The prepared grains are filled in 1/2 liter bottles or polypropylene bags and autoclave for 2 hours at 121°.
- ❖ Grower should see that after sterilization, the material should have a pH value of 7.
- ❖ The bottles are inoculated with grains or bits of agar medium colonized with mycelium, and incubated at 22-24° in dark place.
- ❖ The mycelium completely spreads through the grains in about 2 weeks.

2 Substrate Preparations

- Oyster mushroom can be cultivated on a large number of agro-wastes having cellulose and lignin which helps in more enzyme production of cellulose that is correlated with more yield such as straw of paddy, wheat and ragi, stalk and leaves of maize, millets and cotton, used citronella leaf, sugarcane bag asse, saw dust, jute and cotton waste, dehulled corncobs, pea nut shells, dried grasses, sunflower stalks, used tea leaf waste, discarded waste paper and synthetic compost of button mushrooms etc.
- It can also be cultivated by using industrial wastes like paper mill sludges, coffee byproducts, tobacco waste, apple pomace etc.

8.2.2 Methods of Substrate preparation

The popular methods of substrate preparation are:

- Steam Pasteurization
- Hot Water Treatment;
- Sterile Technique (Till method);
- Fermentation or Composting;
- Chemical Sterilization

i. Steam pasteurization

- In this, pre-wetted straw is packed in wooden trays or boxes and are kept in pasteurization room at 58-620C for 4 hours.
- In this, temperature of pasteurization room is manipulated using steam through a boiler.
- Here, the substrate after cooling at room temperature gets seeded with spawn.
- The complete steam pasteurization process takes atleast 3-5 days.

ii. Hot water treatment

- In this, substrate which can be wheat straw after chopping gets soaked in hot water at around 65-70°C for 1 hrs and at 85°C for 30-45 minutes when using paddy straw .
- Once excess water drains out and cooled, spawn is added.
- Hot water treatment makes the hard substrate such as maize cobs, stems, etc. soft so that growth of mycelia takes place easily.
- Such method is not good for large-scale commercial cultivation.

iii. Chemical sterilization technique

- In this, different species of Trichoderma, Gliocladium, Penicillium, Aspergillus and Doratomycs spp. are competitors fungi on straw during oyster mushroom cultivation.
- If these are there on straw during spawn run, they do not allow growth of mushroom mycelium resulting in yield loss or complete crop failure.

- In this, if wheat straw or paddy straw is treated by steeping in chemical solution of carbendenzim 50% WP and formaldehyde for 16-18 h, then many competitor moulds are killed or growth gets suppressed for 25-40 days after spawning.

iv. Sterile technique

- It is called as Till method where chopped substrate after soaking in cold water is kept in heat resistant polypropylene bags and are sterilized in autoclave at 20 p.s.i. for 1-2 hours with spawning under aseptic conditions.
- Sterile technique is good for research work instead of large-scale commercial production.

v. Fermentation or composting

- It is the advancement of composting technique used for white button mushroom.
- It is good for hard substrates like cotton stalks, maize stalks, leguminous stubbles, etc.
- In this, aerobic and anaerobic fermentation of substrate is suitable for Pleurotus cultivation.
- Here composting to be carried out on covered area or shed.
- For doing this, the substrate gets chopped in 5-6 cm of long pieces and ammonium sulphate or urea and lime of around 0.5%-1% and 1% are added on dry weight basis.
- Grower can also add horse manure or chicken manure of around 10% dry weight instead of nitrogenous fertilizers.
- When lime is added, physical structure of the compost gets improved and after mixing the ingredients, sprinkle water till it gets completely wet.
- Now prepare triangular heap of 75-90 cm with not more than 1 meter height and after 2 days of fermentation, turn the pile by adding 1% superphosphate and 0.5% lime.
- With this, the compost gets ready after 2 days of turning.

3 Spawning of Substrate

- Freshly prepared grain spawn is best for spawning.
- Old spawn of around 3-6 months stored at room temperature 20-30^0 C forms very thick mat like structure due to mycelium aggregation and sometimes young pinheads and fruit bodies start developing in the spawn bottle itself.
- The spawning should be done in a pre-fumigated room (48hrs.with 2% formaldehyde).

8.3 Insect pest and disease management in oyster mushroom

8.3.1 Introduction

- Diseases and pests arise due to lack of diligence and prudence in handling the production of the maintenance process which lead to rise in new job as diseases and pests that attack needs to be addressed.
- Many growers cope up such diseases using fungicides, insecticides and other chemicals, but usage of chemicals turned out to pose new problems, as plant in oyster

mushrooms results in pollution and unhealthy chemicals to be consumed, thereby lowering the selling price.

- Diseases of oyster mushroom usually caused by fungi, molds, bacteria or viruses.
- Oyster mushrooms or baglog the disease is usually characterized by emergence of colored stains, slimy, or physical damage to oyster mushroom fruiting bodies that cannot be harvested.
- In general, the incidence of fungal disease is caused due to lack of sterility production process from nurseries to incubation.

8.3.2 Modes of Diseases spreading

Diseases in oyster mushrooms can spread by following ways / media:

1. air
2. water
3. soil
4. human
5. seed

Pests and diseases such as fungal spores pengkontaminasi, nuisance bacteria, or viruses can spread easily through air flow. Even insect pests can be spread by flying against the flow of air. Similarly, water, soil, humans, and the seeds can carry the same disease as the source of air.

8.3.3 Different Pests

The types of pests and diseases that could interfere with the growth of oyster mushrooms such as insects, spiders, worms, slugs, termites, fungi, parasites and saprophyte, as well as bacteria and viruses.

1. Insect

- Flies and mosquitoes are insects found in kumbung are not properly maintained.
- Insects usually go along with entry and exit of workers, by vents or through small holes that were not detected.
- Humid conditions coupled with aroma of substrate / media log is preferred that these insects eventually multiply in kumbung.
- Insect will lay its eggs on medium baglog and after hatching, the larvae will eat the growing mycelium and fruit body so that oyster mushroom stems with holes and growth of oyster mushroom fruiting bodies gets disturbed.
- Once insect enters the phase of active adult, insects will move to medium mushroom logs that are still healthy and breed, so that within certain period, it can cause considerable damage.
- The insects commonly act as vectors / carriers of disease / virus that could interfere with the growth of oyster mushrooms.

Many types of insect-pests that can transmit diseases such as mildew on kumbung can be:

- Licoriella spp
- lycoriella auripila
- Megaselia spp
- megaselia
- Lepidocyrtus spp
- lepydocertus

Prevention of attacks

- Such insects can be done by placing small wire netting in vents and put a clear plastic on the outside of door to refract light so that insects tend to shy away and stay away from kumbung.
- If this practice is still lacking, it can be done by way of efforts to control insect traps insects in kumbung form of glue that is applied evenly on a sheet of paper / plastic yellow.

2. Spiders

- Spiders can eat mycelium and fruiting bodies of oyster mushroom.
- These spider can spread the fungal spores bullies.

Prevention

- Use chalk powder by spreading on surface of floor and walls kumbung.
- If there are cobwebs, it must be immediately destroyed.

3. Worms

- Worm pest usually eats worms that can cause fungal mycelium which did not grow at all / failure to thrive.
- Worm pests is very small (± 1 mm) and can multiply rapidly.

Prevention

- Worm pest prevention can be done with perfect sterilization process so that the eggs of worm dies.

4. Snail

- Kumbung room is not clean and its floors are dirty and muddy which invite snails.
- Snails will eat the oyster mushroom fruiting bodies grow, so the new growth of oyster mushrooms to be not optimal / damaged.

Prevention

- The natural way to prevent or overcome the slug attack is by spraying the floor and shelves kumbung with extracts of Jatropha.

5. Termite

- ❖ Detecting the presence of termites is relatively difficult.
- ❖ Termites eat the substances contained in wood which is cellulose.
- ❖ Cellulose substance is present in oyster mushroom baglog media so that the possibility of damage to baglog also results quite large.

Prevention

- ❖ Spray the anti-termite chemicals.
- ❖ The use of lemongrass extract to be sprayed onto the soil or affected part of kumbung attack.

8.3.4 Disease management

Several types of diseases commonly found in oyster mushrooms are:

1. *Trichoderma* spp

Trichoderma can be spread through the air or carried away by the workers. The characteristics of contamination caused by this fungus is the emergence of green spots spots or stains on the media so baglog oyster mushroom oyster mushroom mycelium growth becomes stunted. Trichoderma is usually a lot of media found on mushroom logs that have died or on the soil surface. How to solve this problem is to immediately dispose of oyster mushroom log media that has been contaminated. While prevention can be done by performing sterilization / desinfektasi labor and equipment used for the treatment kumbung.

2. *Mucor* spp.

Contamination Mucor marked with black spotting on the surface of baglog media. This contamination led to the growth of Mucor competition with oyster mushroom mycelium. Prevention can be done by reducing the amount of the composition baglog mushrooms and set up / lower the temperature of the room by opening and regulating the circulation of air.

3. *Neurospora* spp

Neurospora can inhibit the growth of mycelium and fruiting bodies. Neurospora raises flour "orange" on the surface of the cotton stopper baglog. Prevention is done by performing sterilization baglog media perfectly and reduce the amount of the composition baglog oyster mushrooms.

4. *Penicillium* spp

Contamination of mycelial growth of Penicillium marked with dark brown / maroon. Prevention can be done by maintaining the cleanliness of the incubation chamber. Meanwhile, in order to cope with Penicillium attack from spreading is to dispose of contaminated media baglog.

8.4 Formulation in compost preparation for milky mushroom

- Milky mushrooms are medium to large in size with rounded caps averaging 10-14 centimeters in diameter that connect to an elongated, thick stem.
- It is botanically classified as Calocybe indica, which are the only mushroom species that is both native to and cultivated in hot, humid climate of India.
- In this, smooth, firm caps are convex when young, flattening out with age, and remain pure white through maturity.
- In Milky mushroom underneath the cap, there are many white, crowded gills, where white stem is about 10 centimeters in height with dense, meaty consistency.
- Milky mushrooms are available with many stems which grows from single base and do not lose their namesake color or discolor with age or handling.
- Milky mushroom grows on wide range of substrates containing lignin, cellulose and hemicelluloses.
- For this, substrate should be fresh and dry as exposed substrate to rain or harvest premature are prone to weed moulds that lead to failure of crop.
- It can be grown on straw of paddy, wheat, ragi, maize/bajra/cotton stalks and leaves, sugarcane bagasse, cotton and jute wastes, dehulled maize cobs, tea/coffee waste etc.
- These are good source of vitamins B2, E, and A, phosphorus, potassium, and selenium which contain calcium, vitamin C, iron, and zinc.
- Milky mushrooms are best for cooked applications like sautéing, steaming, grilling and boiling.
- They can only be grown in regions where temperature is between 75-95 degrees Fahrenheit and where humidity is high.

8.5 Cultivation methods of milky mushroom

8.5.1 Material required for Cultivation

For cultivation of milky mushroom, following materials are required:

Substrate:

- Different types of cellulosic agricultural residues like paddy straw, wheat straw, barley straw, maize, jowar and bajra stalkm groundnut haulms, sugarcane bagasse, wheat bran and cotton waste can be used as basal substrate.

- Paddy straw is best substrate for cultivation of Milky Mushroom as about 1kg of dry straw is required for raising a single bag.

Mushroom Spawn:

- For raising mushroom spawn, 3 weeks - 1month is required for 100 grams of good quality seeds.
- For this, spawn should be procured from recognized spawn laboratory.

Organic Supplement:

- For improving productivity, grower may use pasteurized maize meal, wheat bran, paddy husk or boiled wheat grain at 100-150 gm per bag during spawning.

Polythene Bag:

- Polythene tube of dimension 60 cm x 40 cm with 100 gauge thickness and open at both sides is required for milky mushroom cultivation.

8.5.2 Cultivation Procedure

1. Substrate and substrate preparation

- Milky mushroom can be grown on wide range of substrates as in case of oyster mushroom.
- It can be grown on substrates containing lignin, cellulose and hemicelluloses.

Selection of substrate

- Different agro wastes were used as substrate, which were processed in hot water at 80^0 centigrade for suitable periods as per quality of material.
- In some material, only steaming was done while in others, chemicals were used like for paddy straw for processing.
- As boiling takes long time and does not give good result which makes the process costly, the method of hot water treatment was found to be more suitable.
- Substrate selected should be fresh and dry as substrates exposed to rain or harvested premature are prone to various weed moulds which may result in failure of crop.
- Substrate can be grown on straw of paddy, wheat, ragi, maize/bajra/cotton stalks and leaves, sugarcane bagasse, cotton and jute wastes, dehulled maize cobs, tea/coffee waste etc.,
- Cereal straws are easily available in abundance, which is widely used.

Treatment of Substrate

The substrate can be treated in various ways as follows:

a. Hot water treatment

- In this, water is boiled in container where chopped wet straw filled in gunny bag is submersed in it.

- The main idea of soaking is to saturate the substrate with water as it easy to soak, if straw is filled in gunny bag and dipped in water.
- The water gets boiled for 40 minutes at 80-90°C to get pasteurization.
- It is a popular method particularly for small growers.

b. Steam pasteurization

- The wet straw is filled in insulated room in perforated shelves or in wooden trays.
- In this, steam is released under pressure from boiler while temperature inside the substrate is raised to 65°C and maintained for 5-6 hours.
- Air inside the room needs to be circulated to have uniform temperature in substrate.

c. Chemical sterilization technique

- In this technique, 90 litres of water is taken in rust proof drum or tub having 200 litres capacity.
- In drum or tub, 10-12 kg of wheat straw is slowly steeped in water.
- Take another plastic bucket, and dissolve Bavistin of around 7.5 g and 125 ml of formaldehyde and slowly poured on soaked wheat straw.
- With this, straw is pressed and covered with polythene sheet.
- After 15-18 hour, straw is taken out and excess water gets drained out.
- The chemicals to be added can be calculated as per the quantity while remaining solution can be used at least once for chemical sterilization of straw without any further addition of chemicals.
- Many growers fill the pre-wetted substrate in nylon net bags and then press these bags in cemented tank with chemical solution, as it makes the process of taking out of substrate easy.

d. Sterilization / Autoclaving

- In this, substrate is filled in polypropylene bags having 35x45cm dimensions that can hold around 2-3 kg of wet substrate.
- Now sterilize the substrate at 15 lb psi for 1 hour.
- Once pasteurization/sterilization is over, shift the straw to spawning room for cooling, bag filling and spawning.

2. Spawning and spawn run

- In this, higher spawn dose of around 4-5% of wet substrate is used.
- After spawning bags get shifted to spawn running room and kept in dark where temperature is 25-35°C and relative humidity is above 80% are maintained.
- It takes about 20 days when substrate is fully colonised and bags are ready for casing.

3. Casing

- Casing results in covering of top surface of bags once the spawn run is over using sterilized casing material having thickness of around 2-3 cm.
- It gives physical support, moisture and allow gases to escape from the substrate.
- Casing material such as soil of around 75% and sand of around 25% having pH of around 7.8-7.9 with chalk powder gets sterilized in autoclave at 151b psi for an hour or can be chemically treated with formaldehyde solution of around 4% for a week.
- Solution should be enough to saturate the soil.
- Cover the solution with polythene sheet to avoid escape of chemical and at interval of 2 days, soil to turned so that during casing, soil is free from formalin fumes.
- In this, top of the bag should be made uniform by ruffling top surface of substrate and sprayed with solution of carbendazim and formaldehyde.
- Casing material is spread in uniform layer of 2-3 cm thickness and is sprayed with solution of carbendazim and formaldehyde to he saturation level.
- In this, temperature of around 30-35°C and R.H. of around 80-90% are maintained.

Procedure

- Take full spawn run bed and cut horizontally in equal halves.
- Now compact the beds as much as possible by pressing firmly with hand.
- Apply casing soil to height of 1 cm and then press it slowly.
- Now spray the water sufficiently to wet the cased soil.
- Place the beds inside the blue tent and observe them daily.
- Spray the water, if required, to keep the beds wet.
- Watch for any contamination and insect pests and if the grower found any, then he/she should take necessary steps.
- Now harvest the mushroom, clean it and pack it in polythene bag for sales.
- Stir the top of bed after first harvest and spray water regularly.
- Disturb the topsoil after second harvest and spray water as regular.
- The mushroom yield of 350 –400 g can be obtained from 250 g dry weight of the straw, providing all optimum conditions inside the mushroom shed.

Cropping

- While cropping, mycelium takes 10 days to reach on top of casing layer when there appears fresh air thereby maintaining temperature and R.H.
- For cropping, light should be provided for longer time around 10-12 hour daily.
- The changes made in the environment will result in initiation of fruiting bodies within 3-5 days in form of needle shape that gets mature in 7-10 days.

- ❖ Mushrooms of 7-8 cm diameter can be harvested by twisting, cleaning and packaging in perforated polythene/polypropylene bags for marketing.
- ❖ Mushrooms can also be wrapped in klin film for longer storage.

8.6 Formulation in compost preparation for paddy mushroom

- ❖ Paddy straw mushroom or Volvariella volvacea is famous edible mushroom variety as it has delicious taste and high nutritive value.
- ❖ It is an edible mushroom of topics and subtropics which began to be cultivated in China in 1822.
- ❖ Its cultivation needs environment with high temperature of around 33-35°C with 85-90% of humidity.
- ❖ It is defined as a macro fungus having distinctive fruiting body, which can be seen with naked eyes.
- ❖ It basically grows in rice straw stack in tropical and sub tropical zones, having high temperature and rainy climate.
- ❖ The yield of straw mushroom depends on cultivation methods and compost (growing) medium.
- ❖ To grow, straw alone is not sufficient as composting material due to less nutrients and slow decomposition rates, so these mushrooms presently grows in other materials like cotton waste, sugarcane bagasse, dried banana leaves, oil farm bunch waste etc.
- ❖ There were two types of growing media which were prepared by paddy straw and cotton waste which were both tested in indoor and outdoor systems.
- ❖ In this, the fruiting body formation starts with small clusters of white hyphal aggregates known as primordia which is followed by many morphological stages in development process of fruiting body.

8.7 Cultivation procedure and methods of paddy straw mushroom

8.7.1 Cultivation procedure of paddy straw mushroom

- ❖ Cultivation of paddy straw mushroom is very profitable activity which requires low investment.
- ❖ Commercial production can be started by investing only few thousand which is 5-10 k or less.

Preparing Bed

- Fresh dried straw is most suitable for its cultivation.
- The straw should be tied in bundles of 1.2m long x 25 cm diameter size which should be followed by steeping of bundles in water in cemented tank for 24-48 hours.
- The bundles are taken out and put on cemented floor for few hours for draining out excess water.
- The mushroom beds are prepared on raised bamboo platform in thatched hut.
- Four pre-wetted bundles are placed side by side on the platform, facing all the loose ends on one side, while other four bundles are kept with their tied ends on opposite side.
- The loose ends of all 8 bundles meet and overlap each other in middle.
- Spawn is applied on first layer about 15cm away from the outer edge.
- On top of spawned 1st layer, small quantity of gram dhal powder should be applied along with the spawn.
- On top of spawned 1st layer of 8 bundles, again second layer of 8 bundles is kept in similar fashion and are spawned.
- Now, third layer is laid on top of second layer followed by 4th layer of bundles and spawning.
- All such 32 bundles make a single bed which is can be pressed to remove the entrapped air and make it compact for effective spawn run.
- In order to avoid rapid water loss, beds are covered with polythene sheets or gunny sheets.

Procedure of Paddy Straw Mushroom Cultivation

Material Required:

Following are the materials that are needed for preparing Paddy Straw Mushroom which includes:

Paddy straw:

❖ It is important in cultivation of Volvoriella volvaceae.

❖ It is a by-product of rice plant.

❖ It is known as Paira or Puaal.

❖ In this, rice and straw are separated from each other after harvesting of field crop.

❖ Straw of all paddy sp. can be used for bedding material.

Mushroom Spawn:

❖ Spawn or Mushroom Seed is basic need of whole process.

❖ Spawn of mushroom should always be purchased from trusted manufacturer.

Formalin and Calcium Carbonate:

❖ These are used for degerminating paddy straw.

❖ Two percent of formalin is used in water tank.

❖ Formalin can be purchased online in less quantity for non-professional use.

Transparent Polythene Sheet:

❖ Polythene sheets are used for covering spawned mushroom beds.

❖ It can be purchased from any local vendor.

Water Tank:

❖ The size and number of water tank can be increased according to need.

❖ Standard size tank of 12×4×2 feet should be preferred.

❖ A permanent tank is constructed using cement and bricks.

Preparation of mushroom bed

Mushroom bed works as bedding material for spawn and mushroom. It is a thick layer of paddy straw. Mushroom bed can be prepared by following the steps shown:

Begin in evening.

- ❖ Cut the paddy straw after drying to about 1 – 1.5 feet long pieces.
- ❖ Make 1-2 feet thick layer in square shape.
- ❖ Wrap and tie the straw to make it bundle.

Saturation and de-germination of mushroom bed

- ❖ Keep all the bundles in water tank.
- ❖ The water tank should be stuffed with 3/4 part of bundles.
- ❖ Fill the tank with fresh and clean water.
- ❖ The level of water should always be kept over from last layer of bundle.
- ❖ Pour 2% of formalin in water tank.
- ❖ Allow paddy straw to be saturated for 12-13 hrs.

Spawning of mushroom bed

- ❖ Open the outlet of water tank early morning after 12 hrs.
- ❖ Allow all water to flow out and take all bundles out of tank.
- ❖ Keep all the bundles at equal distance of 1-1.5 feet on clean and raised structure.
- ❖ Loose the bundle and start spawning.

Covering mushroom bed

- ❖ Cover the mushroom bed completely with polythene sheet to create micro environment for proper growth of mycelium.
- ❖ Cover the bed for 4-7 days to allow mycelium to grow in favorable condition.

Care and maintenance

- ❖ Remove the mushroom bed after 4-7 days.
- ❖ Maintain the required level of humidity without damaging mycelium growth.
- ❖ Use hazara to spray water.
- ❖ Use pest control measures to give protection against rats.

Harvesting

- ❖ Mushroom can be harvested after 4-7 days of mycelium growth.
- ❖ It should be harvested while fruiting body is round shaped.
- ❖ 8-10 kg of yield can be achieved from 1 kg of spawn.

8.7.2 Cultivation methods of paddy straw mushroom

There are certain common methods that are used for paddy straw mushroom cultivation such as:

Conventional Method

The different steps involved in this method are as follows:

- ❖ Preparation of paddy straw bundles of 0.75 – 1.0 kg (80-95cm long & 12.16cm wide) preferably from hand threshed paddy.
- ❖ Immersing of bundles in clean water for 12-18 hours in a cemented water tank.
- ❖ Draining out of excess water by placing bundles on raised bamboo platform.
- ❖ Making bed by placing 4 bundles side by side and another four bundles similarly but from the opposite side, forming one layer of eight bundles. The open ends of bundles from opposite sides should overlap in the middle.
- ❖ Forming of second, third & fourth layer by intermittent spawning between first and second, second and third and third and fourth layers.
- ❖ Spawning on entire surface of the layers of the beds at a space of 5cm apart leaving margin of 12-15cm from edges.

- Sprinkling of red gram powder over the spawned surface.
- Using 500 gm spawn and 150 gm of red gram powder for a bed of 30-40 kg of dried paddy straw.
- Pressing of bed from the top and covering with clean plastic sheet for maintaining required humidity (80-85%) and temperature (30-350C).
- Removing of plastic sheet after 7-8 days of spawning and maintaining temperature of 28-320C and relative humidity about 80%.
- Mushroom will start appearing after 4-5 days of sheet removal and will continue for next 20 days.
- After crop harvest the substrate can be used for manure in the field.

Improved Cage-cultivation

- In this, material required for cultivation includes paddy straw 60/Cage bundles, spawn bottle 2/Cage, wooden cage 1 no., 1 drum, polythene sheet 4 meters, binding thread 3 meters, sprayer/rose can 1 no., dithane Z-78/ 1 Pkt. (200 gm), bavistin, malathion 1 bottle (250 ml), dettol/formalin 1 bottle (1/2 liter), dao (Hand 1 No. chopper), thermometer 1 No.
- Select dry, fresh and hand threshed paddy straw free from moulds and leafy portion.
- Make 25 cm long and 10 cm thick bundles @ 60 bundles for each cage (Bed).
- Soak the bundles in boiling water for 20-30 minutes and allow cooling and draining of excess water.
- Disinfect the cage and polythene sheet with 2% Formaline or Dettol solution.
- Arrange ten bundles uniformly in the cage as the bottom layer and put some spawn grains over and inside the bundles.
- Put a second layer of ten bundles over the first and spawn as before.
- Repeat this till six layers of bundles are achieved or till filling of the cage.
- Spray solutions of 0.1% Malathion and 0.2% Dithane Z-78 all over the bed. Cover the whole bed with polythene sheet and bind securely with a binding thread.
- Keep the spawned cage in a room or a shed for mycelial run. A warm place with temperature around 300C is helpful for better mycelial growth.
- Remove the polythene sheet after the mycelial run is completed. Maintain high humidity in and around the bed till pinheads appear.
- Pinheads appear within 10-15 days after spawning. Harvest at the egg stage.
- Continue spraying water for the next flush of mushroom to appear within a week or so.

Outdoor Method

The best place to cultivate paddy straw mushroom outdoor is in shade that are created by trees or creepers, which involves following steps:

- Preparation of raised platform either with sand or bamboo poles or wooden planks or bricks.
- Preparation of bundles of 40cm length and 10 cm width.
- Soaking of bundles in running water or in 2% $CaCO_3$ solution.
- Driving of bamboo pole into the center of each end of the bed.
- Preparation of layer of bundles followed by spawning.
- Laying down of 4 layers of bundles during summer months and 7 layers during rainy season.
- Topping of bed with 20cm deep layer of rice straw followed by covering with polythene sheet.
- Removing of polythene sheet after 4 days and sprinkling of water carefully on 6th day. Spraying of water can be avoided during rainy season.
- Prohibit spraying of water after appearance of the mushroom pinheads.

Indoor Method

The indoor method includes following steps:

i) Substrate:

- Cotton waste is taken as substrate for cultivation of paddy straw mushroom as it contains more cellulose and hemi-cellulose.
- The fine texture of cotton waste helps in retention of moisture, which minimizes water requirement at later stages of cropping, and helps in avoiding damage to mushroom primordia.

ii) Compost preparation:

- In this, substrate is wetted with 1% lime and for cotton waste, square wooden rack is used for holding layer of cotton waste about 30cm deep.
- The workers are used to tread the cotton waste so that it absorbs sufficient quantity of water.
- After first layer is trodden, then other layer is applied and will keep on repeating till the required quantity is trodden.
- In case of paddy straw, pile is made by adding sufficient quantity of water mixed with 1% lime.
- The pile is made with wet cotton waste and left to ferment in the open but under cover during rainy season or extreme cold.
- Here, initial turning is given after 2 days and 5% rice bran is mixed when paddy straw substrate and water is added.
- Nothing is added in cotton waste substrate and piling is formed and left for fermentation for next 2 days.

iii) Bedding and Pasteurization:

- The compost is spread on shelves in rooms or in pasteurization tunnel in suitable thickness.
- The thickness of substrate on shelves varies in different seasons from 5-10 cm thick.
- During summer months, lesser thickness is needed, while higher in winter to preserve moisture and heat.
- After 8-12 hours of compost spreading live stream is introduced using rubber hose of 6cm in diameter.
- A temperature of 62^{0}C is maintained for 2 hours for cotton waste compost while temperature of 65^{0}C is maintained for 6 hrs for paddy straw compost.
- After steaming, shed or room is closed to keep temperature of 50^{0}C for next 24-36 hrs which is followed by natural cooling of substrate.
- The compost is spawned on reaching the temperature near 35^{0}C.

iv) Spawning:

- The compost is spawned with fresh spawn at 1.4% of dry weight or 0.4% of wet weight basis of the compost.
- The pieces of broken spawn are inserted in compost at depth of 2-2.5 cm at distance of 12-15 cm.
- The spawn is covered with displaced compost and bed is covered with thin plastic sheet.
- The room temperature is maintained at 32-34^{0}C during spawn running and compost will colonized in next 4-5 days in cotton waste and 5-6 days in paddy straw compost.

v) Fructification and Crop Management:

- During spawn running period, water and light are not needed but little ventilation is needed.
- By the end of 3-4 days, fluorescent light along with little more ventilation is provided in rooms.
- The plastic sheets are removed on 4-5th day followed by little sprinkling of beds with water.
- The pinheads start appearing on 5th – 6th day of spawning.
- After another 4-5 days, the first flush of mushroom will be ready for harvesting.
- The room conditions required for better fructification are temperature 30^{0}C, relative humidity 80%, fluorescent light and intermittent fresh air.
- The quick growth rate of such mushroom demands more supply of water and oxygen, which are antagonistic to each other in practice.
- Watering of compost is not oftenly recommended as it lowers the temperature, suffocates the tiny primordia and reduces yield.

- Crop management is required for best possible combination of light, temperature, ventilation, relative humidity and compost moisture.

Circular Method

The steps involved in circular method includes:

- Preparing bundles of 1 kg each from paddy straw.
- Soaking of bundles in 2% $CaCO_3$ for 12 hours.
- Winding of water soaked bundles around wooden or cemented poles and then mixing of mushroom spawn at 1.5% on dry weight basis.
- Covering of spawned substrate with thin polythene sheet.
- Maintaining room temperature at 32-34^{0}C, RH 85%, with no light but little air circulation for 5-6 days.
- Removing of polythene sheet with spray of water on beds and lowering of room temperature to 28-32^{0}C, with RH of 80% and fluorescent light with enhancement in fresh air injection.
- Maintaining RH of 80% by fine misting on floor or walls without any direct spray on beds.
- Development of mushroom and harvesting by little lifting and twisting at egg stage.
- Packaging and sealing.

Indigenous Chinese Cultivation

a) Compost preparation

- The wheat straw of 10-15cm long pieces needs to be soaked overnight 1% $CaCO_3$ solution.
- Drain excess water by placing straw on ground.
- Piling of compost and covering with plastic sheet.
- Compost turning after 1 to 2 days interval, when pile reaches temperature at 50^{0}C.
- Filling of compost in 70cm x 35cm x 22cm size frame by putting a layer of compost followed by placing of spawn on four sides of layer along with some wheat bran.
- Keeping second layer on top of first and then spawn and wheat bran are added around the edges.
- The third and fourth layers are added like the first and second layers.

b) Arrangement of bed blocks

- Soil base is raised several centimeters surrounding the base of the frame.
- The blocks are arranged in two rows with a gap of 20-25 cm in between.
- Poplar branches are used to provide roofing on the blocks and are bowed in a shape to form the frame.

- Plastic sheet is spread over the frame, which in turn covered by straw.
- Temperature of around 33-35^0 C is maintained.

c) Harvesting of mushrooms

- Pinheads appear after 4-5 days of spawning.
- Total 9-10 days are taken for first harvest after spawning and initial flush lasts for 3 days accounting around 75% of total mushroom yield.
- The bed blocks are watered with 0.5% $CaCO_3$ and covered again.
- The second flush appears after few days which accounts for rest 25% of total mushroom yield.
- 4-5 crops are harvested each year.

d) Spent compost

- The spent compost is dried and used for producing Pleurotus sajor-caju with BE of 80%.
- After P. sajor-caju cultivation, the spent compost can be used as a good soil conditioner.

8.8 Cultivation methods of shiitake mushroom

8.8.1 Information

- Shiitake mushrooms are second most cultivated mushrooms across the world which is mainly consumed in Asia.
- The mushrooms is brown in color, gills and stalks white which involves supplemented sawdust as initial growing method.
- It is also called black forest mushroom or log mushroom (Lentinula edodes).
- The substrate that is the medium is very important for mushroom cultivation.
- Famous for their meaty texture and earthy, smoky flavor, they're the second most commonly cultivated edible mushrooms, readily available worldwide.
- Shiitake mushrooms are said to have more than 10 times the flavor of white button mushrooms.
- Fresh shiitake mushrooms can be refrigerated in a paper bag up to a week.
- Shiitake mushrooms are biologically different from plant and animal derived foods and are good source of copper.
- It contain riboflavin, niacin, zinc and manganese which contribute to antioxidant properties.
- They contain ergothioniene, which is a bioactive compound that gives potent antioxidant, anti-inflammatory and regenerative properties, and lentinan, which help to fight against bacterial, viral and parasitic infections.

8.8.2 Shiitake Mushroom Production

It is also called black forest mushroom or log mushroom. Traditional method of growing shiitake in Japan is growing it on wood logs but in USA and China they found it easy to grow it on synthetic logs due to environmental conservation.

Synthetic Log Method:

Synthetic bag can be made using substrate placed into plastic bags. Substrate + supplements + buffers

a) Substrate formulations

- Substrates-Sawdust (cured if not from broad leafed tree varieties), cottonseed hulls, coffee hulls, corn cobs, sunflower seed hulls
- PH buffer use calcium carbonate or calcium hydroxide.
- Supplement with brown sugar
- N.B. Keep bags on cemented floor for better results
- Substrate preparation is done same way as oyster.

Incubation stages

- Spawn run or mycelia growth stage- stage during which digestion of food takes place
- Mycelial coat formation – mycelia will thicken and become dense forming a coat on the outer side of the sawdust or substrate. The environmental factors that triggers coat formation include carbon dioxide, where higher the CO2, more will be the coating and better will be the crop.
- Bump formation on the substrate coat: This stage can be called the blister stage, popcorn stage. Too many bumps will increase chances of contamination so reduce the number of bumps by lowering CO2 levels on bags by cutting slits on the polythene bags.
- Browning\pigmentation\coat hardening

Oxygen stage and Diffused light-.

In the presence of oxygen and light, oxidation occurs and the white Mycelia turn reddish brown on the surface of the mycelia while middle of the bag remains white. When one third of bag turns brown, they are ready for fruiting.

Fruiting

Is triggered by changing the environment:

- Light
- Humidity
- Temperature
- Aeration

Ways of triggering fruiting

- ❖ Spraying cold water
- ❖ Physical shaking
- ❖ Soaking in cold water
- ❖ Electrical stimulation

Harvesting

- ❖ Shiitake mushroom will be ready for picking after 8 days. After 14 days you can get another flush. First flush of shiitake mushroom comes 3-4 months of incubation. There can be a total of 4 flushes.
- ❖ After harvesting separate harvested bags from the rest to reverse the environment as it was in beginning

Unit 9

Economics

9.1 Economics of button mushroom cultivation

9.1.1 Information

- Button mushroom cultivation is highly scientific activity which if done in proper manner will give very high returns to grower/entrepreneurs, whether he/she cultivates seasonally or round the year.
- Mushroom cultivation involves investment as per the size of unit/ production targets.
- Before starting the production venture, grower/entrepreneur should have thorough knowledge about the field and should have completed the market survey for sale of produce.
- Expenditure on a mushroom farm can be divided into fixed assets and recurring expenditure.

9.1.2 Mushroom Cultivation Expenses

Fixed Assets

The items permanent in nature and last longer than duration of one crop are covered in this category which includes:

- land,
- building,
- boiler,
- blowers,
- compost handling equipments,
- computers,
- air-conditioning equipments,
- shelves, etc

Variable Cost/Recurring Expenditure

This will includes the expenditure which is involved in production of crop which includes raw materials such as:

- compost ingredients,
- spawn,
- casing soil,
- energy cost,
- pesticides,
- labour charges,
- salary of employees etc.

Mushroom cultivation is labour-intensive job which is mostly carried out on large scale where experienced labour is employed for:

- composting,
- spawning,
- casing of beds
- spraying of beds

In this, the wages of workers will normally be higher as ordinary labour can be employed for picking.

Cultivation of A.bisporus can be done seasonally under natural conditions when the temperature is suitable for its cultivation.

Economics of mushroom plant as per production

1. Infrastructure required

- Land will be required for casing soil dump, road, paths etc, so an approximate land needed for such project should be around 5 hectares

2. Economics of the project
3. Investment in construction of infrastructure
4. Investment on machinery
5. Manpower requirement and wages involved / annum
6. Expenses on raw materials
7. Interest and depreciation
8. Cost of production
9. Total production at 100% capacity utilization
10. Canning operation
11. Sale and profit projections

Calculations on economic aspect of White Button Mushroom cultivation

Particular	Details	Amount
Fixed cost		
Crop Room	size (30 x 17 x 9 ft) 3 tier	25,000
Spray pump	1 unit	1500
Thermo-hygrometer	1 unit	500
Bucket	2 unit	500
Weight Balance	1 unit	500
Total		28000
Variable cost		
Spawn, compost		
casing soil	10 ton Rs. 2000/per ton	20000
Pesticides insecticides & formalin		2000
Electricity, fuel, water charges		1500
Polythene sheet	2000 sq.ft.	2000
Miscellaneous		1500
Total		**27000**
Return		
Total production	1800 kg	
Market rate	@ 60 per kg	108000
Cost of Cultivation		27000
Net Return		81000

9.2 Economics of oyster mushroom cultivation

9.2.1 Information

Cultivation technology of oyster mushroom is simple as it does not require costly infrastructure facilities due to seasonal low cost growing rooms.

Cost related to mushroom production

Cost of cultivation

Cost of cultivation = Total Fixed Cost + Total Variable Cost

Depreciation on fixed capital

The depreciation rates for different farm assets were taken as per 10% per annum.

Net income

Net income = Total revenue – Total expenditure for one crop

9.2.2 Cost Analysis of Oyster Mushroom Production

The cost of production of oyster mushroom involves following expenses:

a Non-Recurring Expenditure which includes cost related to:

- Building
- Sprayer pump
- Galvanized tubs
- Jute Bags
- Nylon ropes

b Recurring Expenditure which includes cost related to:

- Paddy straw bales
- Polythene bags
- Cost of spawn
- Formaldehyde chemical

c Depreciation (Non Recurring Expenditure)

d Revenue

- From mushroom
- By-Product

e Net Profit

9.2.3 Financial analysis in mushroom production

Particulars	**Price (Rs.)**
a Non-Recurring Expenditure	
Building cost	20000
Sprayer pump (1 no.)	1500
Galvanized tubs (2 no.)	600
Jute Bags 25no's @ Rs.10 per bag + Nylon ropes	500
Total	22600
b Recurring Expenditure	
Paddy straw 5 bales @ Rs.300.00	1500
Polythene bags 150 gauge thick 60*45cm @ 100/kg (2kg)	200
Cost of spawn Rs.30/packet (20 no's)	600
Formaldehyde 5 litre @ Rs. 40.00/litre	200
Miscellaneous	1000
Total	3500

Depreciation (Non Recurring Expenditure) @ 10%	2260
Total for one year	2260
Total expenditure	5760

c Revenue

From mushroom 95 kg @ Rs. 200 per kilogram	19000
By-Product	1000
Total Revenue	20000
Net Profit	14240

9.3 Economics of milky mushroom cultivation

9.3.1 Information

The cost of production of milky mushroom cultivation depends on:

- cost of raw material
- yield/unit
- production level
- whole sale price

9.3.2 Financial analysis in mushroom production

Table: Fresh mushroom production in kg

Item	Quantity (kg)	Cost(approx.)
Paddy straw	36500	54,750.00
Spawn	4600	2,30,000.00
Polythene bags	500	38,500.00
PP bags(packing)	500	38,500.00
Casing material		10,000.00
Thread ball		2,000.00
Formaldehyde	50 litres	3,750.00
Bavistin	2.5kg	2,000.00
Melathion/Nuvan	2 litres	1,600.00
Bleaching powder	50kg	1,000.00
Labour (Rs.3,000/month)	3 Nos.(12 months)	1,08,000.00
Water		2,400.00
Electricity		3,400.00
Transport		10,000.00

Rent of building/Shed	12,000.00
Miscellaneous expenditure	12,000.00
Total	5,29,900.00

9.4 Economics of Shitake mushroom cultivation

9.4.1 Information:

Shiitake mushroom grows indoors or outdoors on any deciduous wood that retains its bark for many years. Cultivation of shiitake mushrooms offers farmers and growers, good opportunity to use their lands while turning a profit. The start up costs for cultivation of shiitake mushroom cultivation is low, and high profitability in consecutive years of production. As with any farming venture, viability is achieved with good business planning and attention to strategies which minimize costs and optimize efficiency. When shiitake mushrooms are cultivated outdoors on logs in a managed shade environment, then forest farming practice is initiated.

9.4.2 Requirements to Start Shiitake Cultivation

The following is a detailed description of costs involved in starting small-scale cultivation which includes:

Start-up cost

Items vary with the number of logs in production.

- Logs: Hardwood logs
- Spawn
- Sawdust spawn
- Dowel spawn
- Wax
- Drill bit
- Utilities

9.4.3 Equipments required

- Log drilling stand
- Inoculation tool
- Wax melting pot
- Glass baster to apply wax
- Spring scale (55-lb capacity) to weigh logs
- Extension cord
- Drill: high-speed drill
- Water hose and sprinkler
- Scale (to weigh mushrooms)

- Brackets and bag covers to maintain correct temperature and conditions for log
- Soak tank
- Refrigerator

9.4.4 Harvesting and labor considerations

The following is a description of harvesting and labor considerations you will need to consider when starting a small-scale log operation.

- Harvesting costs
- Packaging and labels: Stamps and boxes
- Advertising or marketing charge
- Transportation
- Picking and storage baskets
- Weighing

Glossary

Absorption: The act of absorbing substances from a solution or a gas mixture by a solid or liquid. Unlike in the case of adsorption the process is going in the whole absorbent.

Acidity – pH: The quality of solutions and damp substrates characterized by the number of hydrogen irons in water.

Agar: A product extracted from algae and used as the nutrient medium for growing sterile cultures.

Agaric bed: A mass of substrate for agaric growing stacked in a special way. It can be stowed on the floor, in boxes, packets, etc.

Agaricales: The order of fungi with gill-hymenophore.

Agaricus bisporus: One of the few fungi species that are subject to cultivation. Moreover, cultivated agarics are more useful, tasty and aromatic than wild ones. The cap of new mushrooms is convex, mature mushrooms have flat caps up to 15 sm. in diameter. The edges are down-fold. The pellicle is usually white, cream or brown (then it has small scales). The underside of small mushrooms caps is covered with soft pellicle hiding pink gills which later grow dark-brown with a hue of violet. When torn the pellicle leaves a dark ring on a stalk. The flesh is sturdy, white turning pink when cut. The fungi have a delicious mushroom flavour.

Air humidity: (relative humidity): The ration of the amount of water vapour actually present in the air to the greatest amount possible at the same temperature, expressed as percentage.

Antagonism: The ability of a fungus to restrain other organism's growth or make it impossible by secreting certain substances in the growing mycelium.

Basidia: Sexual spore-bearing organ of basidial fungi which is mace- and sometimes tuber-like or elongated in form. Its surface can bear up to 4 spores (basidiospores).

Basidiomycetes: The class of fungi producing spores on basidias.

Block: A term standing for pressed and mycelium-sawn substrate (compost) mass packed in perforated plastic film.

Carpophore: The fruiting body of higher fungi (fungi stands for mushrooms here).

Clamp: A compost pile stowed stacked by a special tossing combine. It is about 2 m high and wide and may be more than 40 m long. A clamp are not to be packed too tightly, otherwise there will be no air circulation inside and microorganisms won't grow actively enough.

Compost: (Germ. Kompost): The mixture of straw, dung or hens' excrements and different organic or mineral fertilizers. The main source of nutrition for agaric mycelium.

Conditioning: To condition means to bring into line with set norms and standards. In mushroom-growing the term relates to aeration procedure (e. g. conditionong compost on pasteurizing it or conditioning growth-chamber to set fruiting bodies)

Cover layer: The layer of water-retentive materials mixture which is applied on the substrate to encourage fruiting bodies formation.

Crates: Containers for mushroom picking. They can be for one-time or many-times use. They are made out of plastic or carton.

Dimilin: A preparation for fly control.

Fruiting body: The cap and the stalk of a fungi.

Fungicides: (Lat. fungus + caedre): Chemicals that destroy fungi.

Genotype: The genetic constitution of an organism, the sum-total of all genes localized in its chromosomes. In a broader sense the term denotes the sum-total of all hereditiary traits of an organism.

Gobtage (Fr. Gobtage) (or laying cover-mixture): The procedure of coating mycelium-bearing substrate with a thin layer of cover material where fruiting bodies primordia are to form. Gobtage is practiced in agaric cultures cultivation mostly.

Grain mycelium: The grain inoculated with mycelium. It is used for the extensive dissemination of mycelium in substrate.

Growth-chamber: A room to grow agarics. There are also chambers used to pasteurize compost, germinate mycelium and so on. We single out mono- and polyzonal agaric-growing systems depending on the number of chambers assigned for all the growth-stages from sowing mycelium to cropping.

Heterotrophic: (Greek heteros – another + rophe - nutrition, living on organic matter): The name for organisms that have to live on processed substances that are already assimilated by other organisms. All fungi are heterotrophs.

Hydrometer: An instrument for determining air humidity.

Hymenium: A spore-bearing layer on the underside of a cap. The layer is spread on special outgrowths known as hymenophore. Agaricales (e. g. agarics) hymenophore consists of gills called hymenium gills.

Hymenophore: Special outgrowths under blewits flesh with a spore-bearing layer (hymenium) on their surface.

Hyphae: (sing. Hypha) (from the Greek word hyphe - tissue, web): Thin microscopic uni- and multi-cellular filaments forming fungi thalli (mycelium) and fruiting bodies (carpophores).

Hyphosphere: Microscopic medium surronding hyphae.

Inoculation: (eng. Inoculum): The act or process of introducing mycelium into nutrient substrate.

Insecticides: (lat. insectum + caedre): Chemicals used for combating insects.

Julienne: Mushrooms baked with cheese.

Mesophilic organisms: Organisms that develop at an average temperature (as a rule that's 4-32°C).

Micron: A unit of length equal to one millionth of a meter.

Mould: A general term for saprophytic or parasitic fungi. Usually these are rather extensive colonies varied in form and colour. Moulds colonies are dust-like, velvety or scab-like.

Mycelium: (Lat. mycelium): The thallus of a fungus containing uni- and multi-cellular filaments (hyphae), a spawn.

Mycology: The study of fungi.

Mycosphere: The environment adjoining mycelium.

Parasite: An organism living and feeding on another organism injuring and sometimes even destroying it.

Partial pressure: Gas or steam pressure in a mixture that this gas or steam could exert if its volume equaled that of the whole mixture.

Pasteurization: (after Louis Paster): In mushroom-growing – partial sterilization of substrate at the temperature of 60°C and for a certain exposure period.

Pathogen: (Greek Pathos+genos): A morfibic organism which the causative agent of a disease.

Pesticides: (Lat. pestis+caedre): A general term for chemicals used for pest control.

Pin: A formed mushroom fruit body

Primordial: (Lat. primordium): Rudiments of fruiting bodies. At first they are round-oblong while later emerge a cap and a stalk.

Rhizomorphs: Long lace webs of hyphae. They carry water and nutrients from the mycelium to fruiting bodies and ensure fungi life under unfavourable conditions.

Ring: A rim left from sectional veil on brewits stalk.

Saprophyte: An organism living on the remains of plants and animals turning organic matter into the inorganic one.

Saprophytes: Fungi living on dead organic matter, mainly on the decayed parts of plants. Their feed by decomposing dead plant remains.

Sectional veil: The protective coat of fungi hymenium spanned from the cap edge and a stalk. As a fungi matures it tears to release formed spores.

Sowing mycelium: Sterilized wheat, rye or millet inoculated with mycelium germinated from spores or fungi tissue. It is delivered to mushroom-growers in special plastic bags or bottles.

Spawn: (or mycelium), a typical fungi body found in substrate consisting of very thin interwoven branched web-like filaments (hypha).

Spores: The reproductive cells («seeds») of fungi, bacteria and plants.

Sterilization: (Lat. sterilis):Disinfection, the act of annihilating microorganisms with a high temperature, chemicals or by filtration.

Strain: A group of microporganisms (e. g. fungi) with its own peculiarities and strong points. (like a plant kind or an animal breed)

Stroma: A compact pillow-like mass of mycelium, formed on the compost or cover mixture. It indicates that a mushroom will reproduce asexually.

Substrate: (Lat. substratum): The material on which mycelium grows (this can be a tree, soil, compost, grain, etc.).

Substrate dampness: (absolute humidity): The degree of moisture in the substance expressed in %. The index is calculated according to the formula Humidity = (wet weight – dry weight)/dry weight*100%

Tamping: The process of stowing substrate tightly. Note that substrate is not exactly tamped: e. g. while stowing compost into the substrate is compressed under 100kg/m pressure, i. e. slightly tamped.

The first stage: The first composting stage, during which raw materials are mixed, moistened and processed into the nutrient medium for growing mushrooms.

The second stage: The stage, during which mushroom compost is pasteurized and conditioned.

Thermal probe: A device for measuring the temperature in the compost

Thermophiles: Microscopic organisms growing and reproducing at a rather high temperature (70°C).

Tide: Simultaneous mass mushroom maturing.

Tossing: Tedding a substrate pile while composting it and turning it upside down.

Vegetative reproduction stage: The process of mycelium spreading out in the substrate or topsoil.

Index